I0539964

 # Stop!!!

This program cannot be used effectively without online registration.

Without online registration, you won't have access to practice tests, answer sheets, detailed explanations, automated scoring, personalized study plans.

Before moving forward, please complete your registration using the information given below.

Complete Your Registration Now

Scan the QR Code or Visit

lumoslearning.com/a/tedbooks

Use Access Code: WIQV2G

How to Use the Lumos Program?

Congratulations on choosing the Lumos WFE Program! This isn't just a book. It's a comprehensive program designed to help your child succeed on the state test.

Here is how to use the program:

1 Login to your child's account at *lumoslearning.com* using the details sent to your registered email.

2 Have your child take the assigned online Practice Test 1.

3 After finishing the test, click on the *View Study Plan* button on the home page.

4 For each lesson in the study plan, have your child look at the questions in the printed book and answer them in the online answer sheet.

5 After completing all the lessons in the study plan, have your child take Practice Test 2.

6 View your child's performance by going to the *My Reports* section in the top right corner of the home page.

Table of Contents

Chapter 1
The Number System

DIRECTIONS

Do **NOT** write your answers in this book. **OPEN** the Online Answer Sheet by Scanning the **QR Code** or Visit **lumoslearning.com/a/8m001**

Lesson 1: Rational vs Irrational Numbers

1. Which of the following is an integer?

   Ⓐ -3
   Ⓑ $\frac{1}{4}$
   Ⓒ -12.5
   Ⓓ 0.454545...

2. Which of the following statements is true?

   Ⓐ Every rational number is an integer.
   Ⓑ Every whole number is a rational number.
   Ⓒ Every irrational number is a natural number.
   Ⓓ Every rational number is a whole number.

3. Which of the following accurately describes the square root of 10?

   Ⓐ It is rational.
   Ⓑ It is irrational.
   Ⓒ It is an integer.
   Ⓓ It is a whole number.

4. Complete the following statement: Pi is _____ .

   Ⓐ both real and rational
   Ⓑ real but not rational
   Ⓒ rational but not real
   Ⓓ neither real nor rational

5. Complete the following statement: $\sqrt{7}$ is _____.

   Ⓐ both a real and a rational number
   Ⓑ a real number, but not rational
   Ⓒ a rational number, but not a real number
   Ⓓ neither a real nor a rational number

LumosLearning.com

6. The number 57 belongs to which of the following set(s) of numbers?

Ⓐ N only
Ⓑ N, W, and Z only
Ⓒ N, W, Z, and Q only
Ⓓ All of the following: N, W, Z, Q, and R

7. From the following set: {-√5.7, -9, 0, 5.25, 3i, √16}
Select the answer choice that shows the elements which are Natural numbers.

Ⓐ -√5.7, -9, 0, 5.25, 3i, √16
Ⓑ -√5.7, -9, 0, 5.25, 3i
Ⓒ 3i
Ⓓ Positive square root of 16

8. From the following set: {-√5.7, -9, 0, 5.25, 3i, √16}
Select the answer choice that shows the elements that are Rational numbers.

Ⓐ -√5.7, -9, 0, 5.25, 3i, √16
Ⓑ -9, 0, 5.25, √16
Ⓒ 3i
Ⓓ -√5.7

9. Which of the numbers below is irrational?

Ⓐ √169
Ⓑ √4
Ⓒ √16
Ⓓ √3

10. Write the repeating rational number 0.1515... as a fraction.

Ⓐ $\frac{85}{100}$

Ⓑ $\frac{15}{75}$

Ⓒ $\frac{15}{99}$

Ⓓ $\frac{25}{50}$

11. Write the repeating rational number .112112... as a fraction.

Ⓐ $\dfrac{112}{100}$

Ⓑ $\dfrac{112}{99}$

Ⓒ $\dfrac{112}{999}$

Ⓓ $\dfrac{111}{999}$

12. Which of the following is true of the square root of 2?

Ⓐ It is both real and rational.
Ⓑ It is real but not rational.
Ⓒ It is rational but not real.
Ⓓ It is neither real nor rational.

13. Which of the following sets includes the square root of -25?

Ⓐ R, Z, W, N, and Q
Ⓑ R, W, Q
Ⓒ Z, N
Ⓓ None of the above.

14. Complete the following statement:
The number 6.25 belongs to _____.

Ⓐ R, Q, Z, W, and N
Ⓑ R and Q
Ⓒ R and N
Ⓓ Q and Z

15. Complete the sentence:
Irrational numbers may always be written as _____.

Ⓐ fractions
Ⓑ fractions and as decimals
Ⓒ decimals but not as fractions
Ⓓ neither decimals or fractions

LumosLearning.com

16. Which of the following are rational numbers?

Instruction : Mark all the correct options. More than one option may be correct.

Ⓐ $\dfrac{5}{7}$

Ⓑ $\sqrt{10}$

Ⓒ $\sqrt{25}$

Ⓓ π

17. Mark whether each number is rational or irrational.

	Rational	Irrational
$\sqrt{2}$		
$\dfrac{1}{3}$		
0.575		
$\dfrac{\sqrt{12}}{4}$		

18. Identify the irrational number and circle it.

Ⓐ $\dfrac{5}{7}$

Ⓑ 0.1

Ⓒ $\sqrt{10}$

Did You Check Your Score?

YES

Record your score below:

Score (%): _____

Date: _____

NO

► Scan the QR code or Visit *lumoslearning.com/a/8m001*

► Submit your answers using the *Online Answer Sheet*.

► Get your Scores & Detailed Explanations.

DIRECTIONS

Do **NOT** write your answers in this book. **OPEN** the Online Answer Sheet by Scanning the **QR Code** or Visit **lumoslearning.com/a/8m002**

Chapter 1 → Lesson 2: Approximating Irrational Numbers

1. Between which two whole numbers does $\sqrt{5}$ lie on the number line?

   Ⓐ 1 and 2
   Ⓑ 2 and 3
   Ⓒ 3 and 4
   Ⓓ 4 and 5

2. Between which pairs of rational numbers does $\sqrt{5}$ lie on the number line?

   Ⓐ 2.0 and 2.1
   Ⓑ 2.1 and 2.2
   Ⓒ 2.2 and 2.3
   Ⓓ 2.3 and 2.4

3. Order the following numbers on a number line (least to greatest).

   Ⓐ 1.8, 1.35, 2.5, $\sqrt{5}$
   Ⓑ 1.35, $\sqrt{5}$, 1.8, 2.5
   Ⓒ 1.35, 1.8, $\sqrt{5}$, 2.5
   Ⓓ 1.35, 1.8, 2.5, $\sqrt{5}$

4. If you fill in the _____ in each of the following choices with $\sqrt{7}$, which displays the correct ordering from least to greatest?

   Ⓐ ___, 2.5, 2.63, 2.65
   Ⓑ 2.5, ___, 2.63, 2.65
   Ⓒ 2.5, 2.63, ___, 2.65
   Ⓓ 2.5, 2.63, 2.65, ___

5. Which of the following numbers has the least value?

   Ⓐ $\sqrt{0.6561}$
   Ⓑ 0.8
   Ⓒ 0.8...
   Ⓓ 0.8884

6. Choose the correct order (least to greatest) for the following real numbers.

   Ⓐ $\sqrt{5}, 4\frac{1}{2}, 4.75, 2\sqrt{10}$

   Ⓑ $4\frac{1}{2}, \sqrt{5}, 2\sqrt{10}, 4.75$

   Ⓒ $4\frac{1}{2}, 4.75, \sqrt{5}, 2\sqrt{10}$

   Ⓓ $\sqrt{5}, 2\sqrt{10}, 4\frac{1}{2}, 4.75$

7. Which of the following numbers has the greatest value?

   Ⓐ 0.4...
   Ⓑ 0.444
   Ⓒ $\sqrt{0.4}$
   Ⓓ 0.45

8. Which is the correct order of the following numbers when numbering from least to greatest?

   Ⓐ $\sqrt{0.9}, 0.9, 0.999, 0.9...$
   Ⓑ $0.\overline{9}, \sqrt{0.9}, 0.999, 0.9...$
   Ⓒ $0.9, 0.\overline{9}..., \sqrt{0.9}, 0.999$
   Ⓓ $0.9, 0.9..., 0.\overline{999}, \sqrt{0.9}$

9. Write the following numbers from least to greatest.

   Ⓐ $\sqrt{2}, \pi, 3\frac{7}{8}, \frac{32}{8}$

   Ⓑ $\pi, \sqrt{2}, 3\frac{7}{8}, \frac{32}{8}$

   Ⓒ $3\frac{7}{8}, \pi, \sqrt{2}, \frac{32}{8}$

   Ⓓ $\frac{32}{8}, 3\frac{7}{8}, \pi, \sqrt{2}$

10. If you were to arrange the following numbers on the number line from least to greatest, which one would be last?

   Ⓐ 3.6

   Ⓑ $3\dfrac{7}{12}$

   Ⓒ $\sqrt{12}$

   Ⓓ $3\dfrac{9}{10}$

11. Between which of these pairs of rational numbers does $\sqrt{24}$ lie on the number line?

   Ⓐ 4.79 and $4\dfrac{7}{8}$

   Ⓑ $4\dfrac{7}{8}$ and 5.0

   Ⓒ 4.95 and 5.0

   Ⓓ 4.75 and 4.79

12. Between which two integers does $\sqrt{2}$ lie on the number line?

   Ⓐ 0 and 1
   Ⓑ 1 and 2
   Ⓒ 2 and 3
   Ⓓ 3 and 4

13. Between which pair of rational numbers does $\sqrt{2}$ lie on the number line?

   Ⓐ 1.40 and 1.41
   Ⓑ 1.41 and 1.42
   Ⓒ 1.42 and 1.43
   Ⓓ 1.43 and 1.44

14. Between which pair of consecutive integers on the number line does $\sqrt{3}$ lie?

   Ⓐ 1 and 2
   Ⓑ 2 and 3
   Ⓒ 3 and 4
   Ⓓ 4 and 5

LumosLearning.com

15. Between which of the following pairs of rational numbers on the number line does √3 lie?

    Ⓐ 1.70 and 1.71
    Ⓑ 1.71 and 1.72
    Ⓒ 1.72 and 1.73
    Ⓓ 1.73 and 1.74

Did You Check Your Score?

YES	NO
Record your score below: Score (%): _____ Date: _____	► Scan the QR code or Visit *lumoslearning.com/a/8m002* ► Submit your answers using the *Online Answer Sheet*. ► Get your Scores & Detailed Explanations.

End of The Number System

Chapter 2
Expressions and Equations

14

LumosLearning.com

DIRECTIONS

Do **NOT** write your answers in this book. **OPEN** the Online Answer Sheet by Scanning the **QR Code** or Visit **lumoslearning.com/a/8m003**

Lesson 1: Properties of Exponents

1. Is -5^2 equal to $(-5)^2$?

Ⓐ Yes, because they both equal -25.
Ⓑ Yes, because they both equal -10.
Ⓒ Yes, because they both equal 25.
Ⓓ No, because -5^2 equals -25 and $(-5)^2$ equals 25.

2. $\dfrac{X^6}{X^{-2}} =$

Ⓐ $\dfrac{1}{X^3}$

Ⓑ $\dfrac{1}{X^{12}}$

Ⓒ X^4

Ⓓ X^8

3. Which of the following is equal to 3^{-2} ?

Ⓐ $\dfrac{1}{9}$

Ⓑ -9

Ⓒ 9

Ⓓ $\dfrac{1}{6}$

4. Which of the following is equivalent to $X^{(2-5)}$?

Ⓐ X^3

Ⓑ $X^{\frac{1}{3}}$

Ⓒ $\dfrac{1}{X^3}$

Ⓓ 3^X

5. $1^9 =$

Ⓐ 1
Ⓑ 3
Ⓒ 9
Ⓓ $\dfrac{1}{9}$

6. $(X^{-3})(X^{-3}) =$

Ⓐ X^6
Ⓑ X^9
Ⓒ $\dfrac{1}{X^6}$
Ⓓ $\dfrac{1}{X^9}$

7. $(X^{-2})^{-7} =$

Ⓐ X^5

Ⓑ X^{14}

Ⓒ $\dfrac{1}{X^5}$

Ⓓ $\dfrac{1}{X^{14}}$

8. $(X^4)^0 =$

Ⓐ X
Ⓑ X^4
Ⓒ 1
Ⓓ 0

9. $(3^2)^3 =$

  Ⓐ 3^5
  Ⓑ 3^6
  Ⓒ 3
  Ⓓ 1

10. $5^2 + 5^3 =$ _____

11. Which of the following show the proper laws of exponents?
    Note : More than one option may be correct. Select all the correct answers.

  Ⓐ $3^2 \times 3^5 = 3^{10}$
  Ⓑ $(4^2)^3 = 4^6$
  Ⓒ $\dfrac{8^5}{8^1} = 8^4$
  Ⓓ $7^4 \times 7^4 = 7^8$

12. Simplify this expression. Write your answer in the box below.
    $a^7(a^8)(a)$

    ┌───┐
    │   │
    └───┘

13. Select the ones that properly applied the different Laws of Exponents, making sure to keep positive exponents.
    Note: More than one option may be correct. Select all the correct answers.

  Ⓐ $(4a^3)^2 = 16a^6$
  Ⓑ $(2x^4)^2 = 4x^6$
  Ⓒ $(x^2y^{-1})^2 = \dfrac{x^4}{y^2}$
  Ⓓ $(2a^{-2})^3 = 8a^6$

 **Did You Check Your Score?**

YES	NO
Record your score below:	► Scan the QR code or Visit *lumoslearning.com/a/8m003* 
Score (%): _____	► Submit your answers using the *Online Answer Sheet*.
Date: _____	► Get your Scores & Detailed Explanations.

DIRECTIONS

Do **NOT** write your answers in this book. **OPEN** the Online Answer Sheet by Scanning the **QR Code** or Visit **lumoslearning.com/a/8m004**

Chapter 2 → Lesson 2: Square & Cube Roots

1. **What is the cube root of 1,000 ?**

   Ⓐ 10

   Ⓑ 100

   Ⓒ $33\frac{1}{3}$

   Ⓓ $333\frac{1}{3}$

2. $8\sqrt{12} \div \sqrt{15} =$

   Ⓐ $\dfrac{4}{5}$

   Ⓑ $\dfrac{8}{5}$

   Ⓒ $\dfrac{16\sqrt{5}}{5}$

   Ⓓ $\dfrac{\sqrt{5}}{8}$

3. **The square root of 75 is between which two integers?**

   Ⓐ 8 and 9
   Ⓑ 7 and 8
   Ⓒ 9 and 10
   Ⓓ 6 and 7

4. **The square root of 110 is between which two integers?**

   Ⓐ 10 and 11
   Ⓑ 9 and 10
   Ⓒ 11 and 12
   Ⓓ 8 and 9

5. Solve the following problem: $6\sqrt{20} \div \sqrt{5} =$ _____

    Ⓐ 12
    Ⓑ 11
    Ⓒ 30
    Ⓓ 5

6. The cube root of 66 is between which two integers?

    Ⓐ 4 and 5
    Ⓑ 3 and 4
    Ⓒ 5 and 6
    Ⓓ 6 and 7

7. Which expression has the same value as $3\sqrt{144} \div \sqrt{12}$?

    Ⓐ $3\sqrt{12}$
    Ⓑ $4\sqrt{12}$
    Ⓒ $27 \div \sqrt{12}$
    Ⓓ $33 \div \sqrt{12}$

8. The cubic root of 400 lies between which two numbers?

    Ⓐ 5 and 6
    Ⓑ 6 and 7
    Ⓒ 7 and 8
    Ⓓ 8 and 9

9. Which of the following is equivalent to the expression $4\sqrt{250} \div 5\sqrt{2}$?

    Ⓐ $4\sqrt{25} \div 5$
    Ⓑ $4\sqrt{125} \div \sqrt{2}$
    Ⓒ $4\sqrt{10}$
    Ⓓ $4\sqrt{5}$

10. The cube root of 150 is closest to which of the following?

    Ⓐ 15
    Ⓑ 10
    Ⓒ 5
    Ⓓ 3

11. Select all that apply: What is $\sqrt{\dfrac{81}{289}}$?

Ⓐ $\dfrac{1}{2}$

Ⓑ $\dfrac{9}{17}$

Ⓒ $\dfrac{-1}{2}$

Ⓓ $\dfrac{-9}{17}$

12. Select all the numbers which have integers as the cube roots.

Ⓐ $\sqrt[3]{27}$

Ⓑ $\sqrt[3]{9}$

Ⓒ $\sqrt[3]{1000}$

Ⓓ $\sqrt[3]{18}$

13. Fill in the boxes to make the statement true

$\sqrt[3]{8}$ = ☐ since ☐ × ☐ × ☐ = 8

 **Did You Check Your Score?**

YES	NO
Record your score below:	► Scan the QR code or Visit *lumoslearning.com/a/8m004*
Score (%): _____	► Submit your answers using the *Online Answer Sheet*.
Date: _____	► Get your Scores & Detailed Explanations.

DIRECTIONS

Do **NOT** write your answers in this book. **OPEN** the Online Answer Sheet by Scanning the **QR Code** or Visit **lumoslearning.com/a/8m005**

Chapter 2 → Lesson 3: Scientific Notation

1. In 2007, approximately 3,380,000 people visited the Statue of Liberty. Express this number in scientific notation.

   Ⓐ 0.388×10^7
   Ⓑ 3.38×10^6
   Ⓒ 33.8×10^5
   Ⓓ 338×10^4

2. The average distance from Saturn to the Sun is 890,800,000 miles. Express this number in scientific notation.

   Ⓐ 8908×10^8
   Ⓑ 8908×10^5
   Ⓒ 8.908×10^8
   Ⓓ 8.908×10^5

3. The approximate population of Los Angeles is 3.8×10^6 people. Express this number in standard notation.

   Ⓐ 380,000
   Ⓑ 3,800,000
   Ⓒ 38,000,000
   Ⓓ 380,000,000

4. The approximate population of Kazakhstan is 1.53×10^7 people. Express this number in standard notation.

   Ⓐ 153,000
   Ⓑ 1,530,000
   Ⓒ 15,300,000
   Ⓓ 153,000,000

5. The typical human body contains about 2.5×10^{-3} kilograms of zinc. Express this amount in standard form.

    Ⓐ 0.00025 kilograms
    Ⓑ 0.0025 kilograms
    Ⓒ 0.025 kilograms
    Ⓓ 0.25 kilograms

6. If a number expressed in scientific notation is $N \times 10^5$, how large is the number?

    Ⓐ Between 1,000 (included) and 10,000
    Ⓑ Between 10,000 (included) and 100,000
    Ⓒ Between 100,000 (included) and 1,000,000
    Ⓓ Between 1,000,000 (included) and 10,000,000

7. Red light has a wavelength of 650×10^{-9} meters. Express the wavelength in scientific notation.

    Ⓐ 65.0×10^{-10} meters
    Ⓑ 65.0×10^{-8} meters
    Ⓒ 6.50×10^{-7} meters
    Ⓓ 6.50×10^{-11} meters

8. A strand of hair from a human head is approximately 1×10^{-4} meters thick. What fraction of a meter is this?

    Ⓐ $\dfrac{1}{100}$

    Ⓑ $\dfrac{1}{1,000}$

    Ⓒ $\dfrac{1}{10,000}$

    Ⓓ $\dfrac{1}{100,000}$

9. Which of the following numbers has the greatest value?

    Ⓐ 8.93×10^3
    Ⓑ 8.935×10^2
    Ⓒ 8.935×10^3
    Ⓓ 89.35×10^1

10. Which of the following numbers has the least value?

   Ⓐ -1.56×10^2
   Ⓑ -1.56×10^3
   Ⓒ 1.56×10^2
   Ⓓ 1.56×10^3

11. Which of the following are correctly written in scientific notation?

   Note that more than one option may be correct. Select all the correct options

   Ⓐ $.032 \times 10^5$
   Ⓑ 11.002×10^{-1}
   Ⓒ 1.23×10^5
   Ⓓ 9.625×10^{-7}

12. Change 2,347,000,000 from standard form to scientific notation by filling in the blank boxes.

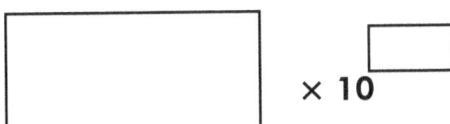

13. Convert 0.0000687 to scientific notation by filling in the blank boxes.

 **Did You Check Your Score?**

YES	NO
Record your score below: Score (%): _____ Date: _____	► Scan the QR code or Visit *lumoslearning.com/a/8m005* ► Submit your answers using the *Online Answer Sheet.* ► Get your Scores & Detailed Explanations.

DIRECTIONS

Do **NOT** write your answers in this book. **OPEN** the Online Answer Sheet by Scanning the **QR Code** or Visit **lumoslearning.com/a/8m006**

Chapter 2 → Lesson 4: Solving Problems Involving Scientific Notation

1. The population of California is approximately 3.7×10^7 people. The land area of California is approximately 1.6×10^5 square miles. Divide the population by the area to find the best estimate of the number of people per square mile in California.

   Ⓐ 24 people
   Ⓑ 240 people
   Ⓒ 2,400 people
   Ⓓ 24,000 people

2. Mercury is approximately 6×10^7 kilometers from the Sun. The speed of light is approximately 3×10^5 kilometers per second. Divide the distance by the speed of light to determine the approximate number of seconds it takes light to travel from the Sun to Mercury.

   Ⓐ 2 seconds
   Ⓑ 20 seconds
   Ⓒ 200 seconds
   Ⓓ 2,000 seconds

3. Simplify $(4 \times 10^6) \times (2 \times 10^3)$ and express the result in scientific notation.

   Ⓐ 8×10^9
   Ⓑ 8×10^{18}
   Ⓒ 6×10^9
   Ⓓ 6×10^{18}

4. Simplify $(2 \times 10^{-3}) \times (3 \times 10^5)$ and express the result in scientific notation.

   Ⓐ 5×10^{-8}
   Ⓑ 5×10^{-15}
   Ⓒ 6×10^8
   Ⓓ 6×10^2

5. Washington is approximately 2.4×10^3 miles from Utah. Mary drives 6×10 miles per hour from Washington to Utah. Divide the distance by the speed to determine the approximate number of hours it takes Mary to travel from Washington to Utah.

    Ⓐ 41 hours
    Ⓑ 40 hours
    Ⓒ 39 hours
    Ⓓ 38 hours

6. Which of the following is NOT equal to $(5 \times 10^5) \times (9 \times 10^{-3})$?

    Ⓐ 4.5×10^4
    Ⓑ 4.5×10^3
    Ⓒ 4,500
    Ⓓ 45×100

7. Find $(5 \times 10^7) \div (10 \times 10^2)$ and express the result in scientific notation.

    Ⓐ 5×10^4
    Ⓑ 0.5×10^5
    Ⓒ 50×10^9
    Ⓓ 5.0×10^9

8. Approximate $.00004567 \times .00001234$ and express the result in scientific notation.

    Ⓐ 5.636×10^{-8}
    Ⓑ 5.636×10^{-9}
    Ⓒ 5.636×10^{-10}
    Ⓓ None of the above.

9. Find the product $(50.67 \times 10^4) \times (12.9 \times 10^3)$ and express the answer in standard notation.

    Ⓐ 653.643
    Ⓑ 65,364,300,000
    Ⓒ 6.53643×10^9
    Ⓓ 6,536,430,000

10 Approximate the quotient and express the answer in standard notation.
    $(1.298 \times 10^4) \div (3.97 \times 10^2)$

    Ⓐ 32.7
    Ⓑ .327
    Ⓒ $.327 \times 10^2$
    Ⓓ None of the above.

11. Select the ones that correctly demonstrate the operations of scientific notation.

Note that more than one option may be correct. Select all the correct answers.

Ⓐ $(4.0 \times 10^3)(5.0 \times 10^5) = 2 \times 10^9$

Ⓑ $\dfrac{4.5 \times 10^5}{9.0 \times 10^9} = 2 \times 10^4$

Ⓒ $(2.1 \times 10^5) + (2.7 \times 10^5) = 4.8 \times 10^5$

Ⓓ $(3.1 \times 10^5) - (2.7 \times 10^2) = 0.4 \times 10^3$

12. Which of the following is correctly ordered from greatest to least?

Note that more than one option may be correct. Select all the correct answers.

Ⓐ 2.0×10^2, 3.0×10^6, 4.0×10^{-7}, 5.0×10^{12}

Ⓑ 4.0×10^{-7}, 2.0×10^2, 3.0×10^6, 5.0×10^{12}

Ⓒ 3.0×10^7, 3.0×10^6, 3.0×10^2, 3.0×10^{-7}

Ⓓ 1.8×10^9, 1.5×10^6, 1.4×10^{-5}, 1.9×10^{-8}

13. $(6 \times 10^3)(9.91 \times 10^0) = $ _____

Did You Check Your Score?

YES	NO
Record your score below: Score (%): _____ Date: _____	► Scan the QR code or Visit *lumoslearning.com/a/8m006* ► Submit your answers using the *Online Answer Sheet*. ► Get your Scores & Detailed Explanations.

DIRECTIONS

Do **NOT** write your answers in this book. **OPEN** the Online Answer Sheet by Scanning the **QR Code** or Visit **lumoslearning.com/a/8m007**

Chapter 2 → Lesson 5: Compare Proportions

1. Find the unit rate if 12 tablets cost $1,440.

   Ⓐ $100
   Ⓑ $150
   Ⓒ $120
   Ⓓ $50

2. A package of Big Bubbles Gum has 10 pieces and sells for $2.90. A package of Fruity Gum has 20 pieces and sells for $6.20. Compare the unit prices.

   Ⓐ Big Bubbles is $0.10 more per piece than Fruity.
   Ⓑ Fruity is $0.02 more per piece than Big Bubbles.
   Ⓒ They both have the same unit price.
   Ⓓ It cannot be determined.

3. The first major ski slope in Vermont has a rise of 9 feet vertically for every 54 feet horizontally. A second ski slope has a rise of 12 feet vertically for every 84 feet horizontally. Which of the following statements is true?

   Ⓐ The first slope is steeper than the second.
   Ⓑ The second slope is steeper than the first.
   Ⓒ Both slopes have the same steepness.
   Ⓓ Cannot be determined from the information given.

4. Which of the following ramps has the steepest slope?

   Ⓐ Ramp A has a vertical rise of 3 feet and a horizontal run of 15 feet
   Ⓑ Ramp B has a vertical rise of 4 feet and a horizontal run of 12 feet
   Ⓒ Ramp C has a vertical rise of 2 feet and a horizontal run of 10 feet
   Ⓓ Ramp D has a vertical rise of 5 feet and a horizontal run of 20 feet

5. Choose the statement that is true about unit rate.

   Ⓐ The unit rate can also be called the rate of change.
   Ⓑ The unit rate can also be called the mode.
   Ⓒ The unit rate can also be called the frequency.
   Ⓓ The unit rate can also be called the median.

6. Which statement is false?

   Ⓐ Unit cost is calculated by dividing the amount of items by the total cost.
   Ⓑ Unit cost is calculated by dividing the total cost by the amount of items.
   Ⓒ Unit cost is the cost of one unit item.
   Ⓓ On similar items, a higher unit cost is not the better price.

7. Selena is preparing for her eighth grade graduation party. She must keep within the budget set by her parents. Which is the best price for her to purchase ice cream?

   Ⓐ $3.99/ 24 oz carton
   Ⓑ $4.80/ one-quart carton
   Ⓒ $11.00 / one gallon tub
   Ⓓ $49.60/ five gallon tub

8. Ben is building a ramp for his skate boarding club. Which of the following provides the least steep ramp?

   Ⓐ 2 feet vertical for every 10 feet horizontal
   Ⓑ 3 feet vertical for every 9 feet horizontal
   Ⓒ 4 feet vertical for every 16 feet horizontal
   Ⓓ 5 feet vertical for every 30 feet horizontal

9. Riley is shopping for tee shirts. Which is the most expensive (based on unit price per shirt)?

   Ⓐ 5 tee shirts for $50.00
   Ⓑ 6 tee shirts for $90.00
   Ⓒ 2 tee shirts for $22.00
   Ⓓ 4 tee shirts for $48.00

10. The swim team is preparing for a meet. Which of the following is Lindy's fastest time?

   Ⓐ five laps in fifteen minutes
   Ⓑ four laps in sixteen minutes
   Ⓒ two laps in ten minutes
   Ⓓ three laps in eighteen minutes

LumosLearning.com

11. David is having a Super Bowl party and he needs bottled sodas. Which of the following purchases will give him the lowest unit cost?

　Ⓐ $2.00 for a 6 pack
　Ⓑ $6.00 for a 24 pack
　Ⓒ $3.60 for a 12 pack
　Ⓓ $10.00 for a 36 pack

12. A package of plain wafers has 20 per pack and sells for $2.40. A package of sugar-free wafers has 30 pieces and sells for $6.30. Compare the unit prices.

　Ⓐ Each plain wafer is $0.17 more than a sugar-free wafer.
　Ⓑ Each sugar-free wafer is $0.09 more than a plain wafer.
　Ⓒ They both have the same unit price per wafer.
　Ⓓ The relationship cannot be determined.

13. Li took 4 practice tests to prepare for his chapter test.
Which of the following is the best score?

　Ⓐ 36 correct out of 40 questions
　Ⓑ 24 correct out of 30 questions
　Ⓒ 17 correct out of 25 questions
　Ⓓ 15 correct out of 20 questions

14. Mel's class is planning a fundraiser. They have decided to have a carnival. If they sell tickets in packs of 40 for $30.00, what is the unit cost?

　Ⓐ $4.00 per ticket
　Ⓑ $0.50 per ticket
　Ⓒ $1.75 per ticket
　Ⓓ $0.75 per ticket

15. Which of the following ski slopes has the steepest slope?

　Ⓐ Ski Slope A has a vertical rise of 4 feet and a horizontal run of 16 feet
　Ⓑ Ski Slope B has a vertical rise of 3 feet and a horizontal run of 12 feet
　Ⓒ Ski Slope C has a vertical rise of 3 feet and a horizontal run of 9 feet
　Ⓓ Ski Slope D has a vertical rise of 5 feet and a horizontal run of 25 feet

16. Solve for the proportion for the missing number.

$$\frac{2}{7} = \frac{4}{\boxed{}}$$

Fill in the blank box with the correct answer.

17. Solve for the proportion for the missing number.

$$\frac{20}{\boxed{}} = \frac{16}{20}$$

Fill in the blank box with the correct answer.

18. Write a proportion to solve this word problem. Then solve the proportion.

Kasey bought 32 kiwi fruit for $16. How many kiwi can Lisa buy if she has $4?

Did You Check Your Score?

YES	NO
Record your score below: Score (%): _____ Date: _____	► Scan the QR code or Visit *lumoslearning.com/a/8m007* ► Submit your answers using the *Online Answer Sheet*. ► Get your Scores & Detailed Explanations.

DIRECTIONS

Do **NOT** write your answers in this book. **OPEN** the Online Answer Sheet by Scanning the **QR Code** or Visit **lumoslearning.com/a/8m008**

Chapter 2 → Lesson 6: Understanding Slope

1. Which of the following statements is true about slope?

    Ⓐ Slopes of straight lines will always be positive numbers.
    Ⓑ The slopes vary between the points on a straight line.
    Ⓒ Slope is determined by dividing the horizontal distance between two points by the corresponding vertical distance.
    Ⓓ Slope is determined by dividing the vertical distance between two points by the corresponding horizontal distance.

2. Which of the following is an equation of the line passing through the points (-1, 4) and (1, 2)?

    Ⓐ y = x - 3
    Ⓑ y = 2x + 2
    Ⓒ y = -2x + 4
    Ⓓ y = -x + 3

3. The graph of which equation has the same slope as the graph of y = 4x + 3?

    Ⓐ y = -2x + 3
    Ⓑ y = 2x - 3
    Ⓒ y = -4x + 2
    Ⓓ y = 4x - 2

4. Which of these lines has the greatest slope?

    Ⓐ $y = \frac{8}{5}x - 7$

    Ⓑ $y = \frac{6}{5}x + 4$

    Ⓒ $y = \frac{7}{5}x + 2$

    Ⓓ $y = \frac{9}{5}x - 3$

5. Which of these lines has the smallest slope?

    Ⓐ $y = \frac{1}{8}x + 7$

    Ⓑ $y = \frac{1}{3}x + 7$

    Ⓒ $y = \frac{1}{4}x - 9$

    Ⓓ $y = \frac{1}{7}x$

6. Fill in the blank with one of the four choices to make the following a true statement. Knowing _____ and the y-intercept is NOT enough for us to write the equation of the line.

    Ⓐ direction
    Ⓑ a point on a given line
    Ⓒ the x-intercept
    Ⓓ the slope

7. A skateboarder is practicing at the city park. He is skating up and down the steepest straight line ramp. If the highest point on the ramp is 30 feet above the ground and the horizontal distance from the base of the ramp to a point directly beneath the upper end is 500 feet, what is the slope of the ramp?

    Ⓐ $\frac{500}{30}$

    Ⓑ $\frac{50}{3}$

    Ⓒ $\frac{3}{50}$

    Ⓓ None of these.

8. If the equation of a line is expressed as $y = \frac{3}{2}x - 9$, what is the slope of the line?

    Ⓐ - 9

    Ⓑ +9

    Ⓒ $\frac{3}{2}$

    Ⓓ $\frac{2}{3}$

9. Which of the following is an equation of the line that passes through the points (0, 5) and (2, 15)?

    Ⓐ y = 5x + 5
    Ⓑ y = 5x + 3
    Ⓒ y = 3x + 5
    Ⓓ y = 5x - 5

10. Which of the following equations has the same slope as the line passing through the points (1, 6) and (3, 10)?

    Ⓐ y = 2x - 9
    Ⓑ y = 5x - 2
    Ⓒ y = 4x - 5
    Ⓓ y = 9x - 6

11. Which of the following equations has the same slope as the line passing through the points (3, 6) and (5, 10)?

    Ⓐ y = 2x -12
    Ⓑ y = 11x - 8
    Ⓒ y = -2x - 9
    Ⓓ y = 3x - 5

12. Which equation has the same slope as y = -5x - 4?

    Ⓐ y = 5x +15
    Ⓑ y = -5x - 11
    Ⓒ y = 5x -19
    Ⓓ y = 5x - 13

13. Find the slope of the line passing through the points (3,3) and (5,5).

    Ⓐ 2
    Ⓑ 1
    Ⓒ 3
    Ⓓ 5

14. Which of the following lines has the steepest slope?

    Ⓐ y = 4x+5
    Ⓑ y =-3x + 5
    Ⓒ y = 3x - 5
    Ⓓ They all have the same slope.

15. Which of the following is the equation of the line passing through the points (0,-3) and (-3,0) ?

Ⓐ y = -3x + 3
Ⓑ y = -3x
Ⓒ y = -x - 3
Ⓓ y = -x

16. Find the slope between the points (-12, -5) and (0, 8).
    Write your answer in the box given below.

17. Find the slope between the points (3, -3) and (12, -2).
    Write your answer in the box given below.

18. Find the slope between the points (1, 2) and (5, -7).
    Write your answer in the box given below.

Did You Check Your Score?

YES	NO
Record your score below:	► Scan the QR code or Visit *lumoslearning.com/a/8m008*
Score (%): _____	► Submit your answers using the *Online Answer Sheet*.
Date: _____	► Get your Scores & Detailed Explanations.

DIRECTIONS

Do **NOT** write your answers in this book. **OPEN** the Online Answer Sheet by Scanning the **QR Code** or Visit **lumoslearning.com/a/8m009**

Chapter 2 → Lesson 7: Solving Linear Equations

1. Which two consecutive odd integers have a sum of 44?

   Ⓐ 21 and 23
   Ⓑ 19 and 21
   Ⓒ 23 and 25
   Ⓓ 17 and 19

2. During each of the first three quarters of the school year, Melissa earned a grade point average of 2.1, 2.9, and 3.1. What does her 4th quarter grade point average need to be in order to raise her grade to a 3.0 cumulative grade point average?

   Ⓐ 3.9
   Ⓑ 4.2
   Ⓒ 2.6
   Ⓓ 3.5

3. Martha is on a trip of 1,924 miles. She has already traveled 490 miles. She has 3 days left on her trip. How many miles does she need to travel each day to complete her trip?

   Ⓐ 450 miles/day
   Ⓑ 464 miles/day
   Ⓒ 478 miles/day
   Ⓓ 492 miles/day

4. Find the solution to the following equation: $3x + 5 = 29$

   Ⓐ $x = 24$
   Ⓑ $x = 11$
   Ⓒ $x = 8$
   Ⓓ $x = 6$

5. Find the solution to the following equation:
   7 - 2x = 13 – 2x

   Ⓐ x = -10
   Ⓑ x = -3
   Ⓒ x = 3
   Ⓓ There is no solution.

6. Find the solution to the following equation: 6x + 1 = 4x - 3

   Ⓐ x = -1
   Ⓑ x = -2
   Ⓒ x = - 0.5
   Ⓓ There is no solution.

7. Find the solution to the following equation:
   2x + 6 + 1 = 7 + 2x

   Ⓐ x = -3
   Ⓑ x = 3
   Ⓒ x = 7
   Ⓓ All real numbers are solutions.

8. Find the solution to the following equation: 8x-1=8x

   Ⓐ x = 7
   Ⓑ x = -8
   Ⓒ x = 8
   Ⓓ There is no solution.

9. Which of the answers is the correct solution to the following equation?
   2x + 5x - 9 = 8x - x - 3 - 6

   Ⓐ x = 3
   Ⓑ x = 7
   Ⓒ x = 9
   Ⓓ All real values for x are correct solutions.

10. Solve the following linear equation for y.

$-y + 7y - 54 = 0$

- Ⓐ $y = 0$
- Ⓑ $y = 1$
- Ⓒ $y = 6$
- Ⓓ $y = 9$

11. Solve each equation for the variable. Select the ones whose values of the variables are the same.

Note that more than one option may be correct. Select all the correct answers.

- Ⓐ $-5m = 25$
- Ⓑ $-10c = -80$
- Ⓒ $-7 + g = -12$
- Ⓓ $12m + 20 = -40$

12. Solve for x: $5x + 20 = -20$.
   x = ?

Write your answer in the box given below.

13. Fill in the missing number to make it true if x = -3.

$7 + 3x = 5x + \underline{\hspace{1cm}}$

Did You Check Your Score?

YES	NO
Record your score below: Score (%): _____ Date: _____	► Scan the QR code or Visit *lumoslearning.com/a/8m009* ► Submit your answers using the *Online Answer Sheet*. ► Get your Scores & Detailed Explanations.

DIRECTIONS

Do **NOT** write your answers in this book. **OPEN** the Online Answer Sheet by Scanning the **QR Code** or Visit **lumoslearning.com/a/8m010**

Chapter 2 → Lesson 8: Solving Linear Equations with Rational Numbers

1. Solve the following linear equation: $\frac{7}{14} = n + \frac{7}{14}n$

   Ⓐ $n = 1\frac{1}{2}$

   Ⓑ $n = 3$

   Ⓒ $n = \frac{1}{3}$

   Ⓓ $n = 1$

2. Find the solution to the following equation: $2(2x - 7) = 14$

   Ⓐ $x = 14$
   Ⓑ $x = 7$
   Ⓒ $x = 1$
   Ⓓ $x = 0$

3. Solve the following equation for x.
   $6x - (2x + 5) = 11$

   Ⓐ $x = -3$
   Ⓑ $x = -4$
   Ⓒ $x = 3$
   Ⓓ $x = 4$

4. $4x + 2(x - 3) = 0$

   Ⓐ $x = 0$
   Ⓑ $x = 1$
   Ⓒ $x = 2$
   Ⓓ All real values for x are correct solutions.

LumosLearning.com

5. Solve the following equation for y.
   3y - 7(y + 5) = y - 35

   Ⓐ y = 0
   Ⓑ y = 1
   Ⓒ y = 2
   Ⓓ All real values for y are correct solutions.

6. Solve the following linear equation: $2(x-5) = \frac{1}{2}(6x+4)$

   Ⓐ x= -12
   Ⓑ x= -9
   Ⓒ x= -4
   Ⓓ There is no solution.

7. Solve the following linear equation for x.

   $3x + 2 + x = \frac{1}{3}(12x + 6)$

   Ⓐ x= -4
   Ⓑ x= 2
   Ⓒ There is no solution.
   Ⓓ All real values for x are correct solutions.

8. $\frac{1}{2}x + \frac{2}{3}x + 5 = \frac{5}{2}x + 6$

   Ⓐ x = $\frac{33}{4}$

   Ⓑ x = $\frac{1}{2}$

   Ⓒ x =- $\frac{3}{4}$

   Ⓓ x =- $\frac{6}{5}$

9. Which of the following could be a correct procedure for solving the equation below?

$2(2x+3) = 3(2x+5)$

Ⓐ $4x+5 = 6x+5$
$-2x+5 = 5$
$-2x = 0$
$x = 0$

Ⓒ $2(5x) = 6x+15$
$10x = 6x+15$
$4x = 15$

$x = \dfrac{15}{4}$

Ⓑ $4x+6 = 6x+15$
$-2x+6 = 15$
$-2x = 9$

$x = -\dfrac{9}{2}$

Ⓓ $4x+6 = 6x+15$
$-2x+6 = 15$
$-2x = 9$

$x = \dfrac{2}{9}$

10. Solve the following linear equation:
$0.64x - 0.15x + 0.08 = 0.09x$

Ⓐ $x = -5$
Ⓑ $x = -0.2$
Ⓒ $x = 5.125$
Ⓓ There is no solution.

11. Select the ones that are correct.

Note that more than one option may be correct.

Ⓐ $w - \dfrac{2}{5} = \dfrac{8}{5}$ so w=2

Ⓑ $\dfrac{-5}{8}$ y=15 so y=24

Ⓒ $0.4x-1.2 = 0.15x+0.8$ so x=8

Ⓓ $\dfrac{x}{6} = -5$ so x=30

12. Select the answer choice with the correct order of how you would solve the equation.

$$\frac{3}{4} + \frac{1}{2}\left(m + \frac{1}{4}\right) = \frac{19}{16}$$

$A \rightarrow \left(m + \frac{5}{8}\right)$ $B \rightarrow \frac{1}{2}\left(m + \frac{1}{4}\right) = \frac{19}{16}$ $C \rightarrow \left(m + \frac{1}{4}\right) = \frac{7}{8}$

Ⓐ **A B C**
Ⓑ **B C A**
Ⓒ **C A B**

13. Solve the following linear equation:

$$\frac{8}{16} = n + \frac{8}{16}\,n$$

Did You Check Your Score?

YES	NO
Record your score below:	▶ Scan the QR code or Visit *lumoslearning.com/a/8m010*
Score (%): _____	▶ Submit your answers using the *Online Answer Sheet*.
Date: _____	▶ Get your Scores & Detailed Explanations.

LumosLearning.com

DIRECTIONS

Do **NOT** write your answers in this book. **OPEN** the Online Answer Sheet by Scanning the **QR Code** or Visit **lumoslearning.com/a/8m011**

Chapter 2 → Lesson 9: Solutions to Systems of Equations

1. Which of the following points is the intersection of the graphs of the lines given by the equations $y = x - 5$ and $y = 2x + 1$?

   Ⓐ $(1, 3)$
   Ⓑ $(-1, -4)$
   Ⓒ $(-2, -3)$
   Ⓓ $(-6, -11)$

2. Which of the following describes the solution set of this system?
   $y = 0.5x + 7$
   $y = 0.5x - 1$

   Ⓐ The solution is $(-2, -3)$ because the graphs of the two equations intersect at that point.
   Ⓑ The solution is $(0.5, 3)$ because the graphs of the two equations intersect at that point.
   Ⓒ There is no solution because the graphs of the two equations are parallel lines.
   Ⓓ There are infinitely many solutions because the graphs of the two equations are the same line.

3. Find the solution to the following system:

   $y = 2(2 - 3x)$
   $y = -3(2x + 3)$

   Ⓐ $x = -1; y = 10$
   Ⓑ $x = -2; y = 24$
   Ⓒ $x = -3; y = 22$
   Ⓓ There is no solution.

4. Use the graph, to find the solution to the following system:

$$\frac{x}{2} + \frac{y}{3} = 2$$

$$3x - 2y = 48$$

Ⓐ x = 8, y = -6
Ⓑ x = 10, y = -9
Ⓒ x = 12, y = -3
Ⓓ x = 16, y = 0

5. Which of the following best describes the relationship between the graphs of the equations in this system?

y = 2x - 6
y = -2x + 6

Ⓐ The lines intersect at the point (0, -3).
Ⓑ The lines intersect at the point (3, 0).
Ⓒ The lines do not intersect because their slopes are opposites and their y-intercepts are opposites.
Ⓓ They are the same line because their slopes are opposites and their y-intercepts are opposites.

6. Solve the system:

2x + 3y = 14
2x - 3y = -10

Ⓐ x = 1, y = 4
Ⓑ x = 2, y = 12
Ⓒ x = 4, y = 2
Ⓓ x = 10, y = 10

7. Solve the system:

x = 13 + 2y
x - 2y = 13

Ⓐ x = 0, y = 13
Ⓑ x = 13, y = 0
Ⓒ There is no solution.
Ⓓ There are infinitely many solutions.

8. Solve the system:

    y = 3x - 7
    x + y = 9

    Ⓐ x = 3, y = 6
    Ⓑ x = 4, y = 5
    Ⓒ x = 5, y = 4
    Ⓓ x = 6, y = 3

9. Solve the system:

    2x + 5y = 12
    2x + 5y = 9

    Ⓐ x = 1, y = 2
    Ⓑ x = 2, y = 1
    Ⓒ There is no solution.
    Ⓓ There are infinitely many solutions.

10. Solve the system:

    -4x + 7y = 26
    4x + 7y = 2

    Ⓐ x = -3, y = 2
    Ⓑ x = 3, y = -2
    Ⓒ x = -2, y = 3
    Ⓓ x = 2, y = -3

11. Randy has to raise $50.00 to repair his bicycle. He is only $1.00 short. He has only $1 and $5 bills. If he has one more $1 bills than $5 bills, how many does he have of each?

    Circle the correct answer choice.

    Ⓐ Ten $1-bills, Nine $5-bills
    Ⓑ Nine $1-bills and Eight $5-bills
    Ⓒ Eight $1-bills and Seven $5-bills
    Ⓓ Seven $1-bills and Six $5-bills

12. Anya is three years older than her brother, Cole. In 11 years, Cole will be twice Anya's current age. Find their current ages.

Circle the correct answer choice.

Ⓐ Anya: 11 years old, Cole: 8 years old
Ⓑ Anya: 10 years old, Cole: 7 years old
Ⓒ Anya: 9 years old, Cole: 6 years old
Ⓓ Anya: 8 years old, Cole: 5 years old

 **Did You Check Your Score?**

YES	NO
Record your score below:	► Scan the QR code or Visit *lumoslearning.com/a/8m011*
Score (%): _____	► Submit your answers using the *Online Answer Sheet*.
Date: _____	► Get your Scores & Detailed Explanations.

DIRECTIONS

Do **NOT** write your answers in this book. **OPEN** the Online Answer Sheet by Scanning the **QR Code** or Visit **lumoslearning.com/a/8m012**

Chapter 2 → Lesson 10: Solving Systems of Equations

1. Find the solution to the following system of equations:
   $13x + 3y = 15$ and $y = 5 - 4x$.

   Ⓐ $x = 0, y = 5$
   Ⓑ $x = 5, y = 0$
   Ⓒ $x = 9, y = -31$
   Ⓓ All real numbers are solutions.

2. Solve the system:
   $y = 2x + 5$
   $y = 3x - 7$

   Ⓐ $x = 12, y = 29$
   Ⓑ $x = 3, y = 11$
   Ⓒ $x = 5, y = -2$
   Ⓓ $x = -1, y = 3$

3. Solve the system:
   $2x + 3y = 14$
   $2x - 3y = -10$

   Ⓐ $x = 1, y = 4$
   Ⓑ $x = 2, y = 12$
   Ⓒ $x = 4, y = 2$
   Ⓓ $x = 10, y = 10$

4. Solve the system:
   $x = 13 + 2y$
   $x - 2y = 13$

   Ⓐ $x = 0, y = 13$
   Ⓑ $x = 13, y = 0$
   Ⓒ There is no solution.
   Ⓓ There are infinitely many solutions.

5. Solve the system:
   2x + 5y = 12
   2x + 5y = 9

   Ⓐ x = 1, y = 2
   Ⓑ x = 2, y = 1
   Ⓒ There is no solution.
   Ⓓ There are infinitely many solutions.

6. Solve the system:
   -4x + 7y = 26
   4x + 7y = 2

   Ⓐ x = -3, y = 2
   Ⓑ x = 3, y = -2
   Ⓒ x = -2, y = 3
   Ⓓ x = 2, y = -3

7. Find the solution to the following system:
   y + 3x = 11
   y - 2x = 1

   Ⓐ x = -5, y = 2
   Ⓑ x = 2, y = 5
   Ⓒ x = -2, y = -5
   Ⓓ x = -2, y = 5

8. Solve the system:
   2x + 4y = 14
   x + 2y = 7

   Ⓐ x = -1, y = 4
   Ⓑ x = 1, y = 3
   Ⓒ There is no solution.
   Ⓓ There are infinitely many solutions.

9. Solve the system:
   3(y - 2x) = 9
   x - 4 = 0

   Ⓐ x = -4, y = -5
   Ⓑ x = 4, y = 11
   Ⓒ There is no solution.
   Ⓓ There are infinitely many solutions.

LumosLearning.com

10. Solve the system:
$$10x = -5(y+2)$$
$$y = 3x-7$$

Ⓐ x = -1, y = 4
Ⓑ x = 2, y = 2
Ⓒ x = 2, y = -2
Ⓓ x = 1, y = -4

11. Which of these will have one solution?
Note that more than one option may be correct. Select all the correct answers.

Ⓐ $y= \frac{3}{4} x+1$

$y=-\frac{1}{2} x-4$

Ⓑ $y=-3x+2$
$3x+y=-4$

Ⓒ $y= \frac{1}{3} x -3$
$2x+y=4$

Ⓓ $-x+2y=-2$
$4y=2x-4$

12. Fill in the table with correct solution for each system of equations.

SYSTEM	SOLUTION
$y = \frac{1}{2}x - 1$ $y = -\frac{1}{4}x - 4$	
$y = 2x + 4$ $y = -3x - 1$	
$y = 4$ $y = 7x - 3$	
$y = -\frac{2}{3}x - 4$ $y = \frac{5}{3}x + 3$	

13. Select the systems that have no solution.
    Note that more than one option may be correct. Select all the correct answers.

Ⓐ $y= -4x+7$
   $y=-3x+3$

Ⓑ $y=\frac{3}{4}x-3$
   $y=\frac{3}{4}x+2$

Ⓒ $y= x-2$
   $y=x+2$

Ⓓ $y=2x+3$
   $4x-2y=8$

Ⓔ $y+2x=-12$
   $y=x+15$

14. Fill in the table with correct solution for each system of equations.

SYSTEM	NUMBER OF SOLUTIONS
$\begin{cases} -x + 2y = 14 \\ x - 2y = -11 \end{cases}$	
$\begin{cases} 2x + 5y = 5 \\ -2x - y = -23 \end{cases}$	
$\begin{cases} y = 3x + 6 \\ -6x+2y =12 \end{cases}$	

Did You Check Your Score?

YES	NO
Record your score below: Score (%): _____ Date: _____	► Scan the QR code or Visit *lumoslearning.com/a/8m012* ► Submit your answers using the *Online Answer Sheet*. ► Get your Scores & Detailed Explanations.

LumosLearning.com

DIRECTIONS

Do **NOT** write your answers in this book. **OPEN** the Online Answer Sheet by Scanning the **QR Code** or Visit **lumoslearning.com/a/8m013**

Chapter 2 → Lesson 11: Systems of Equations in Real-World Problems

1. Jorge and Jillian have cell phones with different service providers. Jorge pays $50 a month and $1 per text message sent. Jillian pays $72 a month and $0.12 per text message sent. How many texts would each of them have to send for their bill to be the same amount at the end of the month?

   Ⓐ 2 texts
   Ⓑ 22 texts
   Ⓒ 25 texts
   Ⓓ 47 texts

2. Mr. Stevens is 63 years older than his grandson, Tom. In 3 years, Mr. Stevens will be four times as old as Tom. How old is Tom?

   Ⓐ 17 years
   Ⓑ 18 years
   Ⓒ 20 years
   Ⓓ 22 years

3. Janet has packed a total of 50 textbooks and workbooks in a box, but she can't remember how many of each are in the box. Each textbook weighs 2 pounds, and each workbook weighs 0.5 pounds, and the total weight of the books in the box is 55 pounds. If t is the number of textbooks and w is the number of workbooks, which of the following systems of equations represents this situation?

   Ⓐ $t + w = 55$
      $2t + 0.5w = 50$

   Ⓑ $2t + w = 50$
      $t + 0.5w = 55$

   Ⓒ $t + w = 50$
      $2t + 0.5w = 55$

   Ⓓ $t + w = 55$
      $2.5(t + w) = 50$

4. Plumber A charges $50 to come to your house, plus $40 per hour of labor. Plumber B charges $75 to come to your house, plus $35 per hour of labor. If y is the total dollar amount charged for x hours of labor, which of the following systems of equations correctly represents this situation?

Ⓐ $y = 50x + 40$
$y = 75x + 35$

Ⓑ $y = 50x + 40$
$y = 35x + 75$

Ⓒ $y = 40x + 50$
$y = 75x + 35$

Ⓓ $y = 40x + 50$
$y = 35x + 75$

5. 10 tacos and 6 drinks cost $19.50. 7 tacos and 5 drinks cost $14.25. If t is the cost of one taco and d is the cost of one drink, which of the following systems of equations represents this situation?

Ⓐ $10t + 6d = 19.50$
$7t + 5d = 14.25$

Ⓑ $6t + 10d = 19.50$
$5t + 7d = 14.25$

Ⓒ $10t + 7t = 19.50$
$6d + 5d = 14.25$

Ⓓ $16(t + d) = 19.50$
$12(t + d) = 14.25$

6. Cindy has $25 saved and earns $12 per week for walking dogs. Mindy has $55 saved and earns $7 per week for watering plants. Cindy and Mindy save all of the money they earn and do not spend any of their savings. After how many weeks will they have the same amount saved? How much money will they have saved?

Ⓐ After 4 weeks, they each will have $83 saved.
Ⓑ After 5 weeks, they each will have $85 saved.
Ⓒ After 6 weeks, they each will have $97 saved.
Ⓓ After 7 weeks, they each will have $104 saved.

7. The seventh and eighth grade classes are raising money for a field trip. The seventh graders are selling calendars for $1.50 each and the eighth graders are selling candy bars for $1.25 each. If they have sold a combined total of 1100 items and each class has the same income, find the number of each item that has been sold.

Ⓐ 400 calendars and 700 candy bars
Ⓑ 700 calendars and 400 candy bars
Ⓒ 500 calendars and 600 candy bars
Ⓓ 600 calendars and 500 candy bars

8. Anya is three years older than her brother, Cole. In 11 years, Cole will be twice Anya's current age. Find their current ages.

Ⓐ Anya: 11 years old
   Cole: 8 years old

Ⓑ Anya: 10 years old
   Cole: 7 years old

Ⓒ Anya: 9 years old
   Cole: 6 years old

Ⓓ Anya: 8 years old
   Cole: 5 years old

9. Tom and his sister both decided to get part-time jobs after school at competing clothing stores. Tom makes $15 an hour and receives $3 in commission for every item he sells. His sister makes $7 an hour and receives $5 in commission for every item she sells. How many items would each of them have to sell to make the same amount of money in an hour?

Ⓐ 1 item
Ⓑ 2 items
Ⓒ 3 items
Ⓓ 4 items

10. Lucia and Jack are training for a marathon. Lucia started the first day by running 2 miles and adds 0.25 mile to her distance every day. Jack started the first day by running 0.5 mile and adds 0.5 mile to his distance every day. If both continue this plan, on what day will Lucia and Jack run the same distance?

Ⓐ Day 3
Ⓑ Day 7
Ⓒ Day 9
Ⓓ Day 12

11. The admission fee at a carnival is $3.00 for children and $5.00 for adults. On the first day 1,500 people enter the fair and $5740 is collected. How many children and how many adults attended the carnival?

Select the correct system and answer. There can be more than one correct answer, choose all applicable ones.

Ⓐ $\begin{cases} 3c + 5a = 1500 \\ c + a = 5740 \end{cases}$

Ⓑ $\begin{cases} 3c + 5a = 5740 \\ c + a = 1500 \end{cases}$

Ⓒ $a = 620, c = 880$

Ⓓ $a = 936, c = 564$

12. Match the correct solution to the corresponding word problem.

	15 and 19	$1.50 and $1.05	$1.75 and $1.60	18 and 31
1. You buy 5 bags of chips and 9 bags of pretzels for $16.95. Later you buy 10 bags of chips and 10 bags of pretzels for $25.50. Find the cost of 1 bag of chips and 1 bag of pretzels.				
2. You empty your coin jar and find 49 coins (all nickels and quarter). The total value of the coins is $8.65. Find the number of nickels and quarters.				
3. Jill bought one hot dog and two soft drinks for a cost of $4.95. Jack bought three hot dogs and one soft drink for a cost of $6.85. Find the cost of one hot dog and one soft drink.				
4. There are a total of 34 lions and hyenas. Each lion eats 4 antelope. Each hyena eats 3 antelope. 117 antelope are eaten. Find the number of lions and hyenas.				

Did You Check Your Score?

YES
Record your score below:
Score (%): _____
Date: _____

NO
► Scan the QR code or Visit *lumoslearning.com/a/8m013*
► Submit your answers using the *Online Answer Sheet*.
► Get your Scores & Detailed Explanations.

End of Expressions and Equations

Chapter 3
Functions

DIRECTIONS

Do **NOT** write your answers in this book. **OPEN** the Online Answer Sheet by Scanning the **QR Code** or Visit **lumoslearning.com/a/8m014**

Lesson 1: Functions

1. Which of the following is NOT a function?

   Ⓐ {(2, 3), (4, 7), (8, 6)}
   Ⓑ {(2, 2), (4, 4), (8, 8)}
   Ⓒ {(2, 3), (4, 3), (8, 3)}
   Ⓓ {(2, 3), (2, 7), (8, 6)}

2. Which of the following tables shows that y is a function of x?

   Ⓐ

x	y
1	4
1	7
4	7

   Ⓑ

x	y
1	7
3	8
4	7

   Ⓒ

x	y
3	2
3	7
4	8

   Ⓓ

x	y
1	7
4	7
4	9

LumosLearning.com

3. If y is a function of x, which of the following CANNOT be true?

   Ⓐ A particular x value is associated with two different y values.
   Ⓑ Two different x values are associated with the same y value.
   Ⓒ Every x value is associated with the same y value.
   Ⓓ Every x value is associated with a different y value.

4. Each of the following graphs consists of two points. Which graph could NOT represent a function?

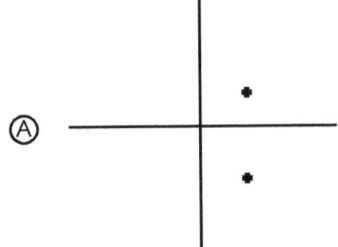

Ⓐ

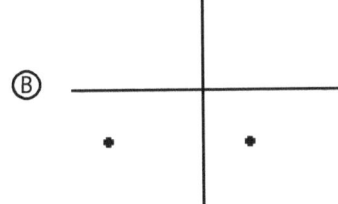

Ⓑ

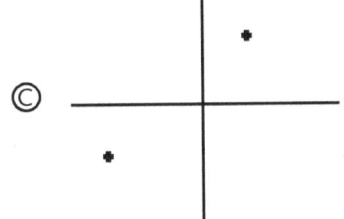

Ⓒ

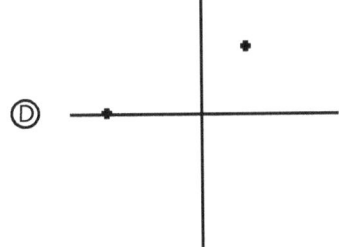

Ⓓ

5. **Which of the following could NOT be the graph of a function?**

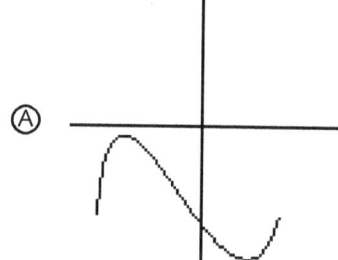

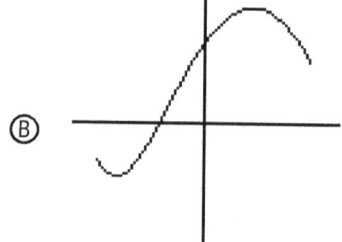

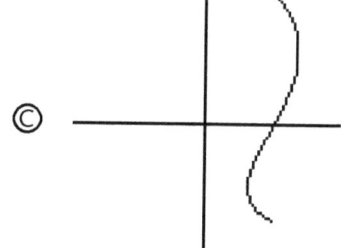

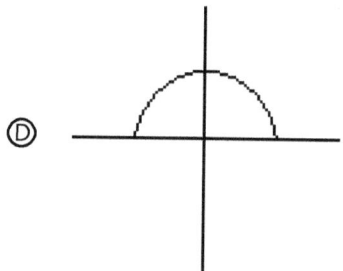

6. **Which of the following is NOT a function?**

Ⓐ {(0, 0), (2, 2), (4, 4)}
Ⓑ {(0, 4), (2, 4), (4, 4)}
Ⓒ {(0, 0), (2, 0), (4, 0)}
Ⓓ {(0, 0), (0, 2), (0, 4)}

LumosLearning.com

7. Which of the following is true of the graph of any non-constant function?

Ⓐ A line drawn parallel to the x-axis will never cross the graph.
Ⓑ A line drawn perpendicular to the x-axis will never cross the graph.
Ⓒ A line drawn through the graph parallel to the x-axis will cross the graph one and only one time.
Ⓓ A line drawn through the graph perpendicular to the x-axis will cross the graph one and only one time.

8. The given set represents a function:
{(0,1), (1,1), (2,1)}
If the ordered pair ____ was added to the set, it would no longer be a function.

Ⓐ (3,1)
Ⓑ (3,2)
Ⓒ (3,3)
Ⓓ (2,3)

9. In order for this set, {(6,5), (5, 4), (4,3)}, to remain a function, which of the following ordered pairs COULD be added to it?

Ⓐ (6,6)
Ⓑ (5,5)
Ⓒ (4,4)
Ⓓ (3,3)

10. Which of the following sets of ordered pairs represents a function?

Ⓐ [(0,1), (0,2), (1,3), (1,4)]
Ⓑ [(1,1), (1,2), (1,3), (1,4)]
Ⓒ [(2,5), (2,6), (4,7), (5,7)]
Ⓓ [(-7,10), (7,10), (8,9), (9,10)]

11. Select the sets of ordered pairs that represent a function.

There can be more than one correct option. Select all the correct options.

Ⓐ {(1,1),(2,2),(3,3),(4,4)}
Ⓑ {(1,-2),(-2,0),(-1,2),(1,3)}
Ⓒ {(-2,3),(0,1),(2,-1),(3,-2)}
Ⓓ {(-1,7),(0,-3),(1,10),(0,7)}

LumosLearning.com

12. Write the function that would go with the table.

x	2	3	4	5
y	-1	0	1	2

13. Select whether or not each example represents a function.

	Function	Not a Function
{(2,10),(2,20),(4,20),(6,30)}	○	○
{(4,10),(8,12),(4,11),(12,13)}	○	○
{(1,3),(2,4),(3,7),(4,13)}	○	○
y=7x-2	○	○

Did You Check Your Score?

YES	NO
Record your score below: Score (%): _____ Date: _____	► Scan the QR code or Visit *lumoslearning.com/a/8m014* ► Submit your answers using the *Online Answer Sheet*. ► Get your Scores & Detailed Explanations.

DIRECTIONS

Do **NOT** write your answers in this book. **OPEN** the Online Answer Sheet by Scanning the **QR Code** or Visit **lumoslearning.com/a/8m015**

Chapter 3 → Lesson 2: Compare Functions

1. A set of instructions says to subtract 5 from a number n and then double that result, calling the final result p. Which function rule represents this set of instructions?

   Ⓐ p = 2(n – 5)
   Ⓑ p = 2n – 5
   Ⓒ n = 2(p – 5)
   Ⓓ n = 2p – 5

2. Which of the following linear functions is represented by the (x, y) pairs shown in the table below?

x	y
-3	-1
1	7
4	13

   Ⓐ y = x + 2
   Ⓑ y = 2x + 5
   Ⓒ y = 3x + 1
   Ⓓ y = 4x + 3

3.

x	y
0	3
1	5
2	7

   Three (x, y) pairs of a linear function are shown in the table above. Which of the following functions has the same slope as the function shown in the table?

   Ⓐ y = 3x + 2
   Ⓑ y = 2x -4
   Ⓒ y = x + 3
   Ⓓ y = x - 1

4. The graph of linear function A passes through the point (5, 6). The graph of linear function B passes through the point (6, 7). The two graphs intersect at the point (2, 5). Which of the following statements is true?

Ⓐ Function A has the greater slope.
Ⓑ Both functions have the same slope.
Ⓒ Function B has the greater slope.
Ⓓ No relationship between the slopes of the lines can be determined from this information.

5. If line M includes the points (-1, 4) and (7, 9) and line N includes the points (5, 2) and (-3, -3), which of the following describes the relationship between M and N?

Ⓐ They are perpendicular.
Ⓑ They intersect but are not perpendicular.
Ⓒ They are parallel.
Ⓓ Not enough information is provided.

6. If line R includes the points (-2, -2) and (6, 4) and line S includes the points(0,4) and (3,0) which of the following describes the relationship between R and S?

Ⓐ They are perpendicular.
Ⓑ They intersect but are not perpendicular.
Ⓒ They are parallel.
Ⓓ Not enough information is provided.

7.

x	y
-4	13
2	1
6	-7

Three points of a linear function are shown in the table above. What is the y-intercept of this function?

Ⓐ (0, 5)
Ⓑ (0, 6)
Ⓒ (0, 7)
Ⓓ (0, 8)

8. Comparing the two linear functions, $y = 2x + 7$ and $y = 7x + 2$, we find that

Ⓐ $y = 2x + 7$ has the steeper slope.
Ⓑ The graphs of the functions will be parallel lines.
Ⓒ The graphs of the functions will intersect at the point (1, 9).
Ⓓ The graphs of the functions will be perpendicular lines.

9. Which of the following is true of the graphs of the lines $y = 3x + 5$ and $y = -\dfrac{x}{3} + 6$?

Ⓐ The graphs will be two parallel straight lines.
Ⓑ The graphs will be two perpendicular straight lines.
Ⓒ Both functions will produce the same line.
Ⓓ There is not enough information to compare.

10. If line A, denoted by $y = x + 9$, and line B, denoted by $y = 5x - 3$ are graphed, which of the following statements is correct?

Ⓐ Line B has a steeper slope.
Ⓑ Both lines have the same slope.
Ⓒ Line A has the steeper slope.
Ⓓ There is not enough information to answer the question.

11. **A function may be described in 3 ways : (1) Set of ordered pairs (x,y) is given or (2) An equation showing how y depends on x or (3) a table showing how y varies with x. In the problem given below, take x to be the time and y to be the distance travelled.**

    Select the ones that have the correct conclusion. Note that more than one option may be correct.

Ⓐ **CJ' s Speed:**

X	Y
0	0
2	13
4	26

Holly's Speed: y=6x Conclusion: Holly is faster than CJ.

Ⓑ **Hank's Speed:**$y= \dfrac{7}{2} x$ **Smith's Speed:** {(0,0),(4,5),(8,10)} Conclusion: Hank is faster than Smith.

Ⓒ **Tyler's Speed:**

X	Y
0	0
4	6
8	12

Omar's Speed:$y= \dfrac{3}{2} x$ Conclusion: Tyler and Omar run the same speed.

12. **Put the correct inequality or equality sign between the rates of change of the two functions given below.**

    Instruction : Take f to be the first function {(0, 0), (3, 180), (6, 360), (9, 540)} and g to be the second function y = 100x

    {(0, 0), (3, 180), (6, 360), (9, 540)} [＿＿＿＿] y = 100x

13. Match each function to whether the rate of change is positive or negative

	Positive	Negative
$y=3x+4$	○	○
$y=\frac{1}{2}x-6$	○	○
$y=-7x+2$	○	○
$y=\frac{2}{3}x-1$	○	○

Did You Check Your Score?

YES	NO
Record your score below: Score (%): _____ Date: _____	► Scan the QR code or Visit *lumoslearning.com/a/8m015* ► Submit your answers using the *Online Answer Sheet*. ► Get your Scores & Detailed Explanations.

DIRECTIONS

Do **NOT** write your answers in this book. **OPEN** the Online Answer Sheet by Scanning the **QR Code** or Visit **lumoslearning.com/a/8m016**

Chapter 3 → Lesson 3: Linear Functions

1. A linear function includes the ordered pairs (2, 5), (6, 7), and (k, 11). What is the value of k?

   Ⓐ 8
   Ⓑ 10
   Ⓒ 12
   Ⓓ 14

2. Which of the following functions is NOT linear?

   Ⓐ f(x) = x + 0.5
   Ⓑ f(x) = -x + 0.5
   Ⓒ f(x) = x² + 0.5
   Ⓓ f(x) = 0.5x

3. Which of the following functions is linear and includes the point (3, 0)?

   Ⓐ f(x) = 3/x
   Ⓑ f(x) = x - 3
   Ⓒ f(x) = 3
   Ⓓ f(x) = 3x

4. Four (x, y) pairs of a certain function are shown in the table below. Which of the following describes the function?

x	y
-3	1
-1	4
1	7
3	10

   Ⓐ The function increases linearly.
   Ⓑ The function decreases linearly.
   Ⓒ The function is constant.
   Ⓓ The function is not linear.

LumosLearning.com

5. Four (x, y) pairs of a certain function are shown in the table below. Which of the following statements describes the function correctly?

x	y
0	3
1	4
2	7
3	12

Ⓐ The function is linear because it does not include the point (0, 0).
Ⓑ The function is linear because it does not have the same slope between different pairs of points.
Ⓒ The function is nonlinear because it does not include the point (0, 0).
Ⓓ The function is nonlinear because it does not have the same slope between different pairs of points.

6. The graph of a linear function lies in the first and fourth quadrants. Which of the following CANNOT be true?

Ⓐ It is an increasing function.
Ⓑ It is a constant function.
Ⓒ It also lies in the second quadrant.
Ⓓ It also lies in the third quadrant.

7. A linear function includes the ordered pairs (0, 1), (3, 3), and (9, n). What is the value of n?

Ⓐ 5
Ⓑ 6
Ⓒ 7
Ⓓ 8

8. A linear function includes the ordered pairs (0, 3), (3, 9), and (9, n). What is the value of n?

Ⓐ 20
Ⓑ 14
Ⓒ 21
Ⓓ 22

9. The graph of a linear function with a non-negative slope lies in the first and second quadrants. Which of the following CANNOT be true?

Ⓐ It is an increasing function.
Ⓑ It is a constant function.
Ⓒ It also lies in the third quadrant.
Ⓓ It also lies in the fourth quadrant.

10. Which function is represented by this table?

x	y
0	2
1	4
2	6
3	8

Ⓐ $y = 2x + 2$
Ⓑ $y = 3x - 4$
Ⓒ $y = 4x - 5$
Ⓓ $y = 6x - 8$

11. Which of the following functions are linear? Select all the correct answers.

Ⓐ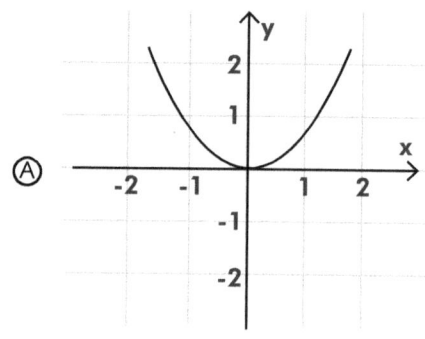

Ⓑ $y = 5x+9$

x	y
-2	-4
-1	-1
0	2
1	5
2	8

©

difference of y-values

$-1 + 4 = 3$

$2 + 1 = 3$

$5 - 2 = 3$

$8 - 5 = 3$

Ⓓ $y = x^2 + 6$

12. Write whether this represents a linear or non-linear function.

$2x^2 + 3y = 10$

13. Match each function to linear or non-linear.

<table>
<tr><th></th><th>Linear</th><th>Non-Linear</th></tr>
<tr><td>$y = x^2$</td><td>○</td><td>○</td></tr>
<tr><td>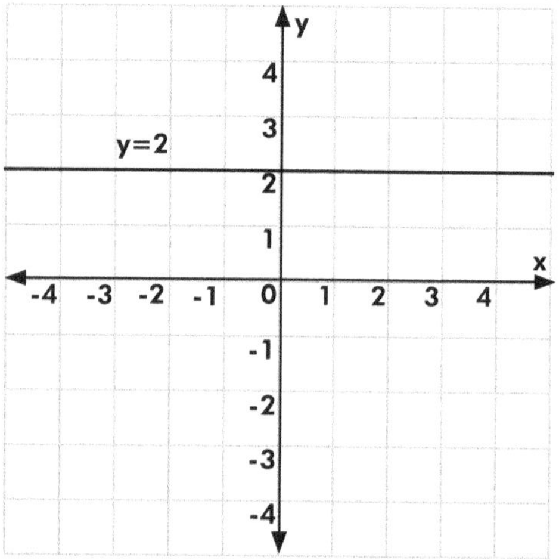
</td><td>○</td><td>○</td></tr>
<tr><td><table>
<tr><th>x</th><th>y</th></tr>
<tr><td>1</td><td>1</td></tr>
<tr><td>2</td><td>8</td></tr>
<tr><td>3</td><td>27</td></tr>
<tr><td>4</td><td>64</td></tr>
<tr><td>5</td><td>125</td></tr>
<tr><td>6</td><td>216</td></tr>
<tr><td>7</td><td>?</td></tr>
</table></td><td>○</td><td>○</td></tr>
<tr><td>$y=-4x+12$</td><td>○</td><td>○</td></tr>
</table>

Did You Check Your Score?

YES

Record your score below:

Score (%): ___________

Date: ___________

NO

- Scan the QR code or Visit ***lumoslearning.com/a/8m016***
- Submit your answers using the *Online Answer Sheet.*
- Get your Scores & Detailed Explanations.

DIRECTIONS

Do **NOT** write your answers in this book. **OPEN** the Online Answer Sheet by Scanning the **QR Code** or Visit **lumoslearning.com/a/8m017**

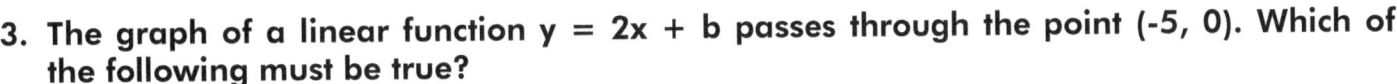

Chapter 3 → Lesson 4: Linear Function Models

1. If a graph includes the points (2, 5) and (8, 5), which of the following must be true?

   Ⓐ It is the graph of a linear function.
   Ⓑ It is the graph of an increasing function.
   Ⓒ It is not the graph of a function.
   Ⓓ None of the above

2. The graph of a linear function y = mx + 2 goes through the point (4, 0). Which of the following must be true?

   Ⓐ m is negative.
   Ⓑ m = 0
   Ⓒ m is positive
   Ⓓ Cannot be determined.

3. The graph of a linear function y = 2x + b passes through the point (-5, 0). Which of the following must be true?

   Ⓐ b is positive.
   Ⓑ b is negative.
   Ⓒ b = 0
   Ⓓ Cannot be determined.

4. The graph of a linear function y = mx + b goes through the point (0, 0). Which of the following must be true?

   Ⓐ m is positive.
   Ⓑ m is negative.
   Ⓒ m = 0
   Ⓓ b = 0

5. The graph of a linear function y = mx + b includes the points (2, 5) and (9, 5). Which of the following gives the correct values of m and b?

   Ⓐ m = 5 and b = 7
   Ⓑ m = 1 and b = 0
   Ⓒ m = 0 and b = 5
   Ⓓ m = 2 and b = 9

6. What is the slope of the linear function represented by the (x, y) pairs shown in the table below?

x	y
0	11
2	7
3	5

   Ⓐ $-\dfrac{1}{2}$

   Ⓑ $\dfrac{1}{2}$

   Ⓒ - 2

   Ⓓ 2

7. The graph of a certain linear function includes the points (-4, 1) and (5, 1). Which of the following statements describes the function accurately?

   Ⓐ It is an increasing linear function.
   Ⓑ It is a decreasing linear function.
   Ⓒ It is a constant function.
   Ⓓ It is a nonlinear function.

8. **Which of the following linear functions has the greatest slope?**

Ⓐ
x	y
0	1
2	5
4	9

Ⓑ
x	y
0	3
2	6
4	9

Ⓒ
x	y
0	5
2	7
4	9

Ⓓ
x	y
0	7
2	8
4	9

9. **A young child is building a tower of blocks on top of a bench. The bench is 18 inches high, and each block is 3 inches high. Which of the following functions correctly relates the total height of the tower (including the bench) h, in inches, to the number of blocks b?**

   Ⓐ h = 3b - 18
   Ⓑ h = 3b + 18
   Ⓒ h = 18b - 3
   Ⓓ h = 18b + 3

10. **Jim owes his parents $10. Each week, his parents pay him $5 for doing chores. Assuming that Jim does not earn money from any other source and does not spend any of his money. Which of the following functions correctly relates the total amount of money m, in dollars, that Jim will have to the number of weeks w?**

   Ⓐ m = -5w - 10
   Ⓑ m = -5w + 10
   Ⓒ m = 5w - 10
   Ⓓ m = 5w + 10

11. An amusement park charges $5 admission and an additional $2 per ride. Which of the following functions correctly relates the total amount paid p, in dollars, to the number of rides, r?

Ⓐ p = 2r + 5
Ⓑ p = 5r + 2
Ⓒ p = 10r
Ⓓ p = 7r

12. Which of the following linear functions has the smallest slope?

Ⓐ
x	y
1	2
4	6
7	8

Ⓒ
x	y
1	4
4	6
7	8

Ⓑ
x	y
1	3
4	6
7	9

Ⓓ
x	y
1	5
4	6
7	7

13. The graph of a linear function includes the points (6, 2) and (9, 4). What is the y intercept of the graph?

Ⓐ (0, 0)
Ⓑ (0, -2)
Ⓒ (3, 0)
Ⓓ (0, 3)

14. Which of the following functions has this set of points as solutions?
{(0, -5), (1, 0), and (4, 15)}

Ⓐ f(x) = 4x - 15
Ⓑ f(x) = 0
Ⓒ f(x) = -5
Ⓓ f(x) = 5x - 5

LumosLearning.com

15. A music store is offering a special on CDs. The cost is $20.00 for the first CD and $10.00 for each additional CD purchased.
Which of the following functions represents the total amount in dollars of your purchase where x is the number of CDs purchased?

Ⓐ f(x) = 10x + 20
Ⓑ f(x) = 10(x - 1) + 20
Ⓒ f(x) = 20x + 10
Ⓓ f(x) = 20(x - 1) + 10

16. A trainer for a professional football team keeps track of the amount of water players consume throughout practice. The trainer observes that the amount of water consumed is a linear function of the temperature on a given day. The trainer finds that when it is 90°F the players consume about 220 gallons of water, and when it is 76°F the players consume about 178 gallons of water. Fill in the value of m in the function : y = mx - 50.

y = _____ x - 50

17. Match each slope and coordinates of point with the correct equation

	y=−4x−3	y=7x-5	y=-x+2
(1, 2) slope=7	○	○	○
(-2, 5) slope=-4	○	○	○
(3, -1) slope=-1	○	○	○

18. A cell phone company charges $89.99 for a new phone and then $19.99 per month. What is the slope or rate of change for this situation?

Slope = _____

19. Plot the (x,y) points shown in the below table on a graph such that it represents a linear function.

x	y
0	11
2	7
3	5

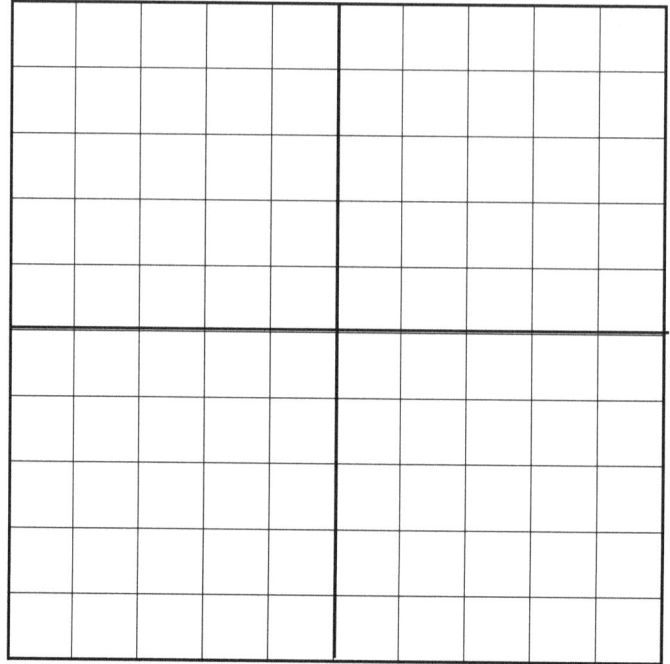

Did You Check Your Score?

YES	NO
Record your score below: Score (%): _____ Date: _____	► Scan the QR code or Visit *lumoslearning.com/a/8m017* ► Submit your answers using the *Online Answer Sheet*. ► Get your Scores & Detailed Explanations.

DIRECTIONS

Do **NOT** write your answers in this book. **OPEN** the Online Answer Sheet by Scanning the **QR Code** or Visit **lumoslearning.com/a/8m018**

Chapter 3 → Lesson 5: Analyzing Functions

1. Complete the following:
   The cost per copy is a function of the number of copies of any one title purchased.
   This implies that _____

   Ⓐ the cost per copy of any one title is always a constant.
   Ⓑ the cost per copy of any one title will change based on the number of copies purchased.
   Ⓒ the cost per copy of any one title is not related to the number of copies purchased.
   Ⓓ None of the above.

2. If a student's math grade is a positive function of the number of hours he spends preparing for a test, which of the following is correct?

   Ⓐ The more he studies, the lower his grade.
   Ⓑ The more he studies, the higher his grade.
   Ⓒ There is no relation between how much he studies and his grade.
   Ⓓ The faster he finishes his work, the higher his grade will be.

3. Mandy took a math quiz and received an initial score of i. She retook the quiz several times and, with each attempt, doubled her previous score.
   After a **TOTAL** of four attempts, her final score was _____.

   Ⓐ 2i
   Ⓑ 3i
   Ⓒ 2^3i
   Ⓓ None of the above.

4. The graph of a linear function lies in the first and third quadrants. Which of the following must be true?

   Ⓐ It also lies in the second quadrant.
   Ⓑ It also lies in the fourth quadrant.
   Ⓒ It is an increasing function.
   Ⓓ It is a decreasing function.

5. The graph of an increasing linear function crosses the vertical axis at (0, -1). Which of the following CANNOT be true?

   (A) It also lies in the second quadrant.
   (B) It also lies in the fourth quadrant.
   (C) It is an increasing function.
   (D) y intercept is -1.

6. A linear equation is plotted on the coordinate plane, and its graph is perpendicular to the x-axis. Which of the following best describes the slope of this line?

   (A) Zero
   (B) Undefined
   (C) Negative
   (D) Positive

7. The graph of a decreasing linear function crosses the vertical axis at (0, 3). Which of the following CANNOT be true?

   (A) The graph lies in the first quadrant.
   (B) The graph lies in the second quadrant.
   (C) The graph lies in the third quadrant.
   (D) The graph lies in the fourth quadrant.

8. Which of the following best describes the x & y coordinates of any point in the first quadrant?

   (A) Both are positive
   (B) Both are negative
   (C) One is positive and one is negative
   (D) None of these

9. In the coordinate plane, in which quadrant is the ordered pair, (-3, -6) located?

   (A) I
   (B) II
   (C) III
   (D) IV

10. In the coordinate plane, in which quadrant is the ordered pair, (1, 8) located?

   (A) I
   (B) II
   (C) III
   (D) IV

11. In the coordinate plane, in which quadrant is the ordered pair, (6, -9) located?

Ⓐ I
Ⓑ II
Ⓒ III
Ⓓ IV

12. A linear equation is plotted on the coordinate plane, and its graph is parallel to the x-axis. Which of the following best describes the slope of this line?

Ⓐ Zero
Ⓑ Undefined
Ⓒ Negative
Ⓓ Positive

13. Which of the following best describes the x & y coordinates of any point in the third quadrant?

Ⓐ Both are positive
Ⓑ Both are negative
Ⓒ One is positive and one is negative
Ⓓ None of these

14. In a coordinate plane the graphs of two functions are perpendicular. Which of the following best describes the relationship of their slopes?

Ⓐ Their slopes are the same.
Ⓑ Their slopes are both zero.
Ⓒ Their slopes are both one.
Ⓓ One slope is the negative reciprocal of the other.

15. In a coordinate plane the graphs of two linear functions are parallel lines. Which of the following best describes the relationship of their slopes?

Ⓐ Their slopes are the same.
Ⓑ Their slopes are opposites.
Ⓒ The slope of one line is the negative reciprocal of the other.
Ⓓ There is no known relationship between their slopes.

16. Observe the graph given.
    Match each segment to whether it is increasing or decreasing as per the graph.

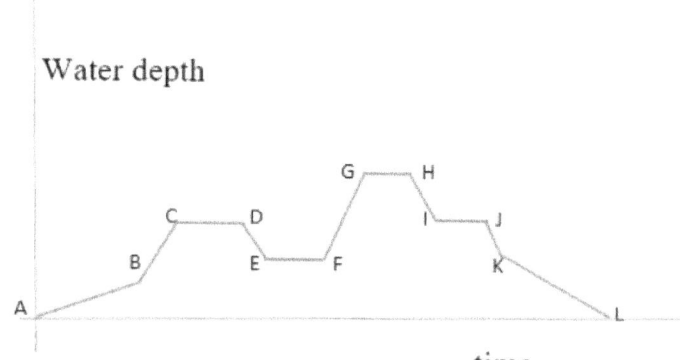

	INCREASING	DECREASING	CONSTANT
A to B	O	O	O
B to C	O	O	O
G to H	O	O	O
I to J	O	O	O
J to K	O	O	O
K to L	O	O	O
C to D	O	O	O
E to F	O	O	O
D to E	O	O	O
F to G	O	O	O
H to I	O	O	O

✋ **Did You Check Your Score?**

YES	NO
Record your score below: Score (%): _____ Date: _____	► Scan the QR code or Visit *lumoslearning.com/a/8m018* ► Submit your answers using the *Online Answer Sheet*. ► Get your Scores & Detailed Explanations.

End of Fraction

Chapter 4

Geometry

DIRECTIONS

Do **NOT** write your answers in this book. **OPEN** the Online Answer Sheet by Scanning the **QR Code** or Visit **lumoslearning.com/a/8m019**

Lesson 1: Transformations of Points & Lines

1. The point (4, 3) is rotated 90° clockwise about the origin. What are the coordinates of the resulting point?

   Ⓐ (-3, 4)
   Ⓑ (-4, 3)
   Ⓒ (4, -3)
   Ⓓ (3, -4)

2. A line segment has a length of 9 units. After a certain transformation is applied to the segment, the new segment has a length of 9 units. What was the transformation?

   Ⓐ A rotation
   Ⓑ A reflection
   Ⓒ A translation
   Ⓓ Any of the above transformations.

3. The point (2, 4) is rotated 180° clockwise about the origin. What are the coordinates of the resulting point?

   Ⓐ (-2, -4)
   Ⓑ (-2, 4)
   Ⓒ (2, -4)
   Ⓓ (2, 4)

4. Two points are located in the (x, y) plane on the opposite sides of the y-axis. After a certain transformation is applied to both points, the two new points end up again on the opposite sides of the y-axis. What was the transformation?

   Ⓐ A rotation
   Ⓑ A reflection
   Ⓒ A translation
   Ⓓ A dilation

5. A certain transformation is applied to a line segment. The new segment shifted to the left within the coordinate plane. What was the transformation?

Ⓐ A rotation
Ⓑ A reflection
Ⓒ A translation
Ⓓ It cannot be determined.

6. A line segment with end points (1, 1) and (5, 5) is moved and the new end points are now (1, 5) and (5,1). Which transformation took place?

Ⓐ reflection
Ⓑ rotation
Ⓒ translation
Ⓓ dilation

7. A certain transformation moves a line segment as follows: A (2, 1) moves to A' (2, -1) and B (5, 3) to B' (5, -3).
Name this transformation.

Ⓐ Rotation
Ⓑ Translation
Ⓒ Reflection
Ⓓ Dilation

8. After a certain transformation is applied to point (x, y), it moves to (y, -x).
Name the transformation.

Ⓐ Rotation
Ⓑ Translation
Ⓒ Reflection
Ⓓ Dilation

9. A transformation moves the point (0, y) to a new location at (0, -y).
Name this transformation.

Ⓐ Rotation
Ⓑ Translation
Ⓒ Reflection
Ⓓ It could be any one of the three listed.

10. (x,y) is in Quadrant 1. Reflection across the y-axis would move it to a point with the following coordinates.

Ⓐ (x, y)
Ⓑ (-x, -y)
Ⓒ (x, -y)
Ⓓ (-x, y)

11. Mark TRUE or FALSE based on the description of the transformation.

	TRUE	FALSE
If you graph a point A (3,2). The point gets translated 10 units down, it will end up at A'(3,12)	○	○
Point A (-2,4) is reflected over the y-axis. The new ordered pair will be A'(2,4).	○	○
Line AB is 3 units long. After it is rotated 90° counter-clockwise, the line will now be 3 units long.	○	○
Point A(5,8) is translated 3 units to the right. It is now located at A'(8,8)	○	○

12. Enter the correct operation that will describe the rule for the translation left 3 units and up 4 units?

(x,y) --> (x ⬚ 3, y ⬚ 4)

13. Circle the graph that represents a rotation.

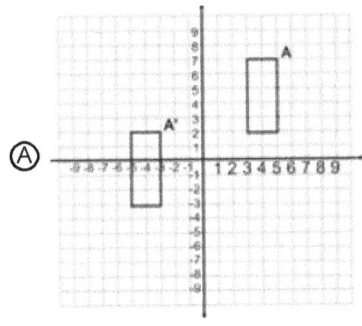

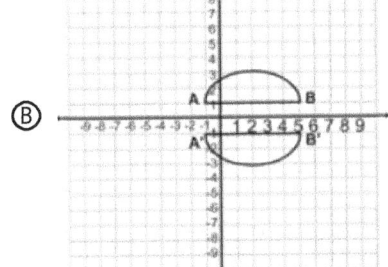

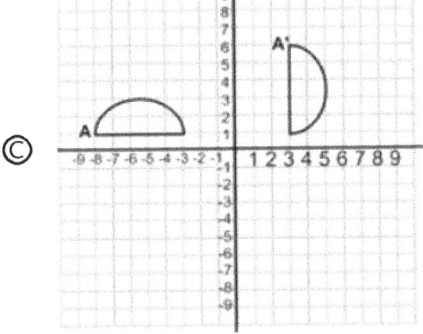

Did You Check Your Score?

YES	NO
Record your score below:	► Scan the QR code or Visit *lumoslearning.com/a/8m019*
Score (%): _____	► Submit your answers using the *Online Answer Sheet*.
Date: _____	► Get your Scores & Detailed Explanations.

DIRECTIONS

Do **NOT** write your answers in this book. **OPEN** the Online Answer Sheet by Scanning the **QR Code** or Visit **lumoslearning.com/a/8m020**

Chapter 4 → Lesson 2: Transformations of Angles

1. **△ABC** is reflected across the x-axis.
   Which two angles are congruent?

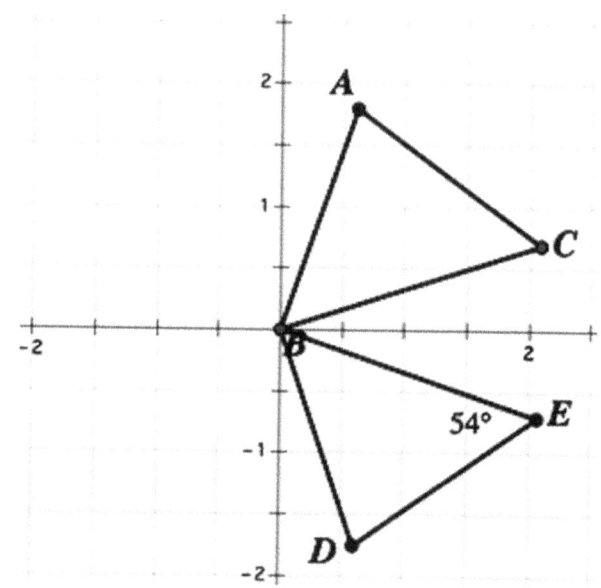

   Ⓐ ∠A and ∠C
   Ⓑ ∠A and ∠E
   Ⓒ ∠C and ∠D
   Ⓓ ∠C and ∠E

2. **△ABC** is rotated 90°.
   Which two angles are congruent?

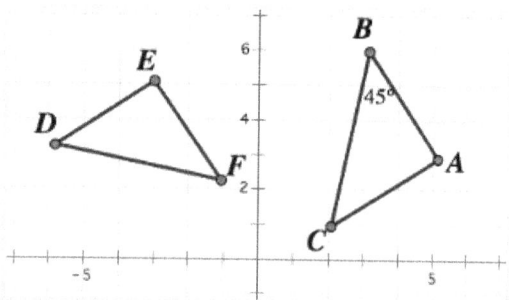

   Ⓐ ∠A and ∠C
   Ⓑ ∠B and ∠E
   Ⓒ ∠C and ∠D
   Ⓓ ∠C and ∠F

LumosLearning.com

3. What rigid transformation should be used to prove ∠**ABC** ≅ ∠**XYZ** ?

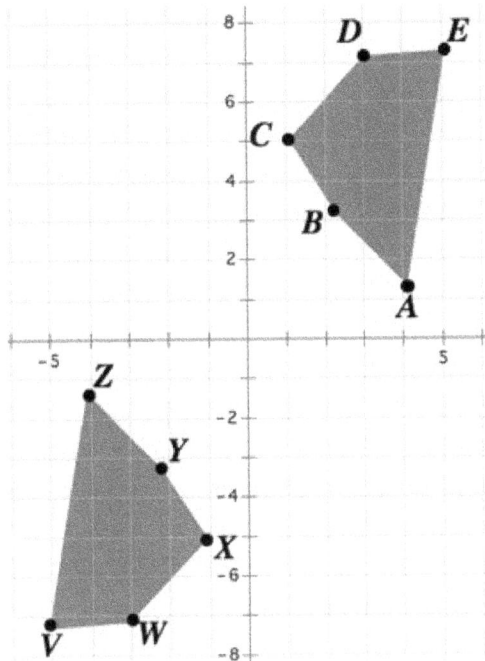

Ⓐ Reflection
Ⓑ Rotation
Ⓒ Translation
Ⓓ None of the above

4. What rigid transformation should be used to prove ∠**ABC** ≅ ∠**XYZ**?

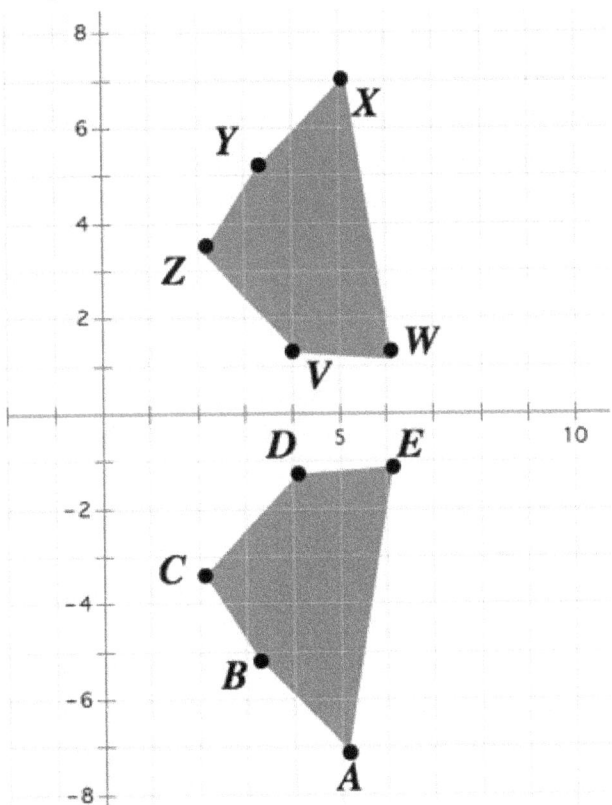

Ⓐ Reflection
Ⓑ Rotation
Ⓒ Translation
Ⓓ None of the above

5. △**ABC** is rotated 90°.
   Which two angles are congruent?

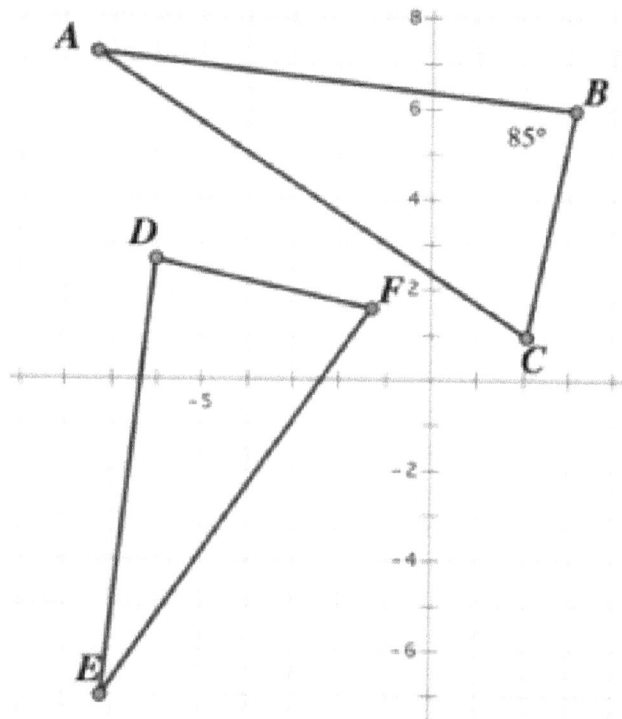

    Ⓐ ∠ A and∠ F
    Ⓑ ∠ B and∠ D
    Ⓒ ∠ B and∠ F
    Ⓓ ∠ C and∠ D

6. If all the triangles below are the result of one or more rigid transformations, which of the following MUST be true?

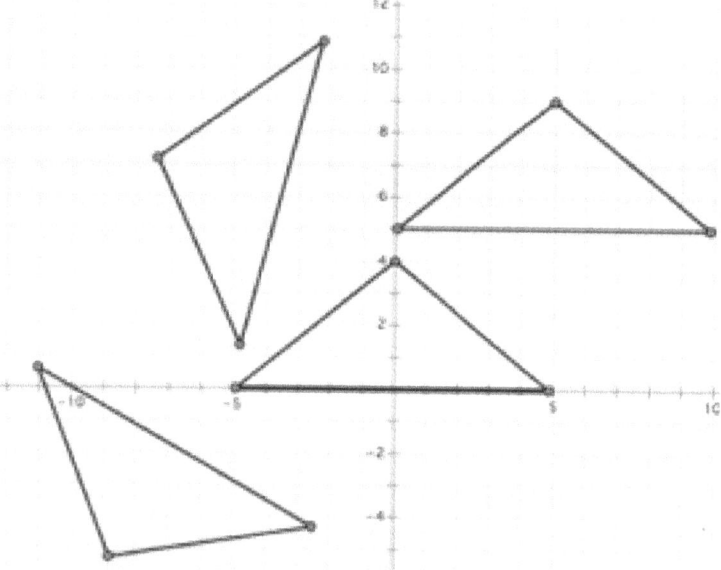

    Ⓐ All corresponding line segments are congruent.
    Ⓑ All corresponding angles are congruent.
    Ⓒ All triangles have the same area.
    Ⓓ A,B, and C are all correct.

7. What rigid transformation should be used to prove ∠*ABC* ≅ ∠*DEF* ?

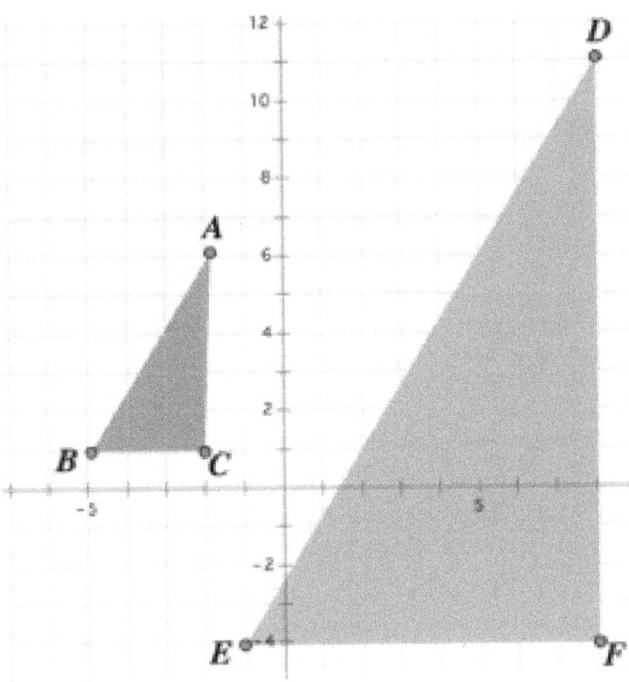

  Ⓐ **Reflection**
  Ⓑ **Rotation**
  Ⓒ **Translation**
  Ⓓ **None of the above**

8. **A company is looking to design a new logo, which consists only of transformations of the angle below:**

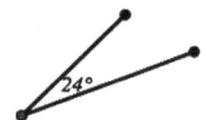

Which logo meets the company's demand?

(A)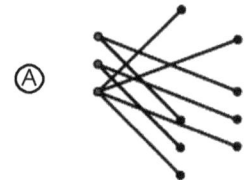

(B)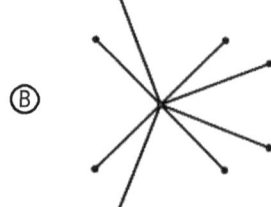

(C)

(D) **All of the above**

9. The angle ∠ AOB is 45° and has been rotated 120° around point C. What is the measure of the new angle ∠ XYZ?

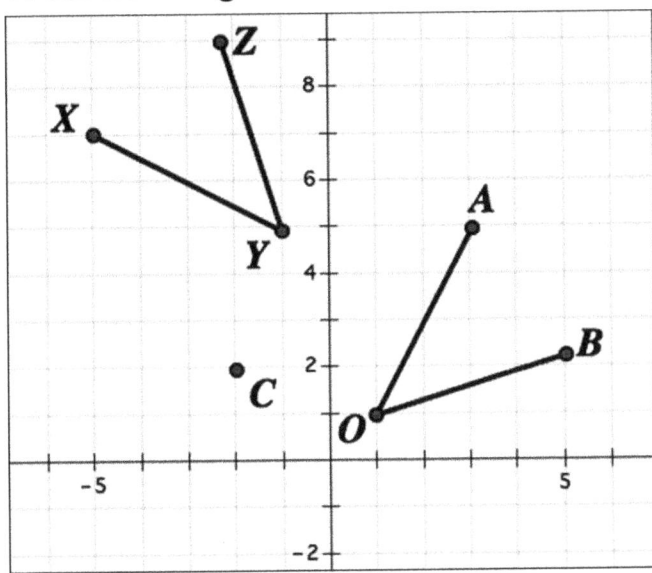

Ⓐ 30°
Ⓑ 45°
Ⓒ 90°
Ⓓ 120°

10. Find the measure of ∠ ABC

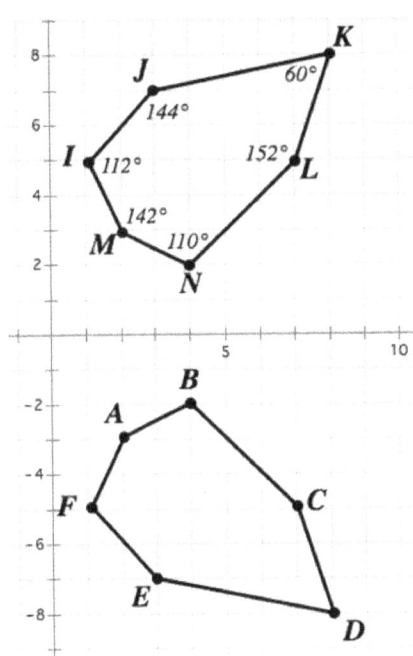

Ⓐ 110°
Ⓑ 112°
Ⓒ 142°
Ⓓ 144°

11. Select the angle measure that corresponds with each transformation. Your preimage has an angle measure of 30°

	30°	60°	90°
Translation	○	○	○
Reflection	○	○	○
Rotation	○	○	○
Dilation	○	○	○

12. In the figure below $\triangle ABC \cong \triangle DEF$. Which angle will correspond with angle B? Type the letter (in capitals) which corresponds to the required angle in the box.

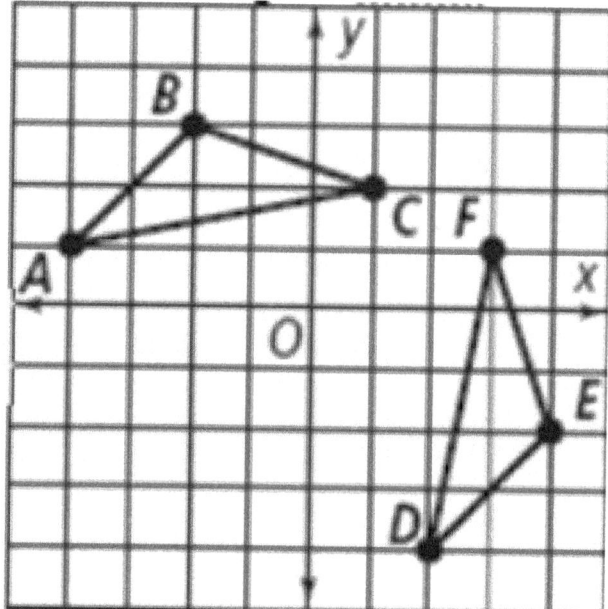

```



```

13 . Triangle ABC is rotated 90° counterclockwise. Which two angles would be congruent?

Circle the correct answer choice.

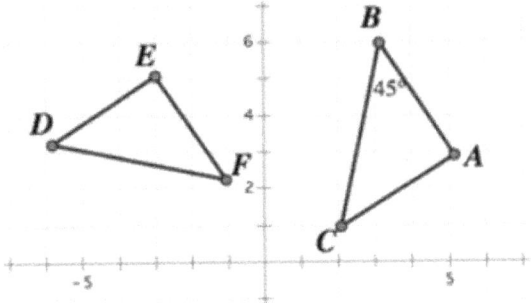

- Ⓐ **Angle A and Angle C**
- Ⓑ **Angle B and Angle E**
- Ⓒ **Angle C and Angle D**
- Ⓓ **Angle C and Angle F**

Did You Check Your Score?

YES	NO
Record your score below:	► Scan the QR code or Visit *lumoslearning.com/a/8m020*
Score (%): _____	► Submit your answers using the *Online Answer Sheet*.
Date: _____	► Get your Scores & Detailed Explanations.

DIRECTIONS

Do **NOT** write your answers in this book. **OPEN** the Online Answer Sheet by Scanning the **QR Code** or Visit **lumoslearning.com/a/8m021**

Chapter 4 → Lesson 3: Transformations of Parallel Lines

1. Two parallel line segments move from Quadrant One to Quadrant Four. Their slopes do not change. What transformation has taken place?

   Ⓐ Reflection
   Ⓑ Translation
   Ⓒ Dilation
   Ⓓ This is not a transformation.

2. Two parallel line segments move from Quadrant One to Quadrant Four. Their slopes change from a positive slope to a negative slope. What transformation has taken place?

   Ⓐ Reflection
   Ⓑ Rotation
   Ⓒ Translation
   Ⓓ It could be either a rotation or a reflection.

3. Two parallel line segments move from Quadrant One to Quadrant Two. Their slopes change from a negative slope to a positive slope. What transformation has taken place?

   Ⓐ Reflection
   Ⓑ Rotation
   Ⓒ Translation
   Ⓓ It could be either a rotation or a reflection.

4. Line *L* is translated along segment $\overline{AB}$ to create line *L′* . Will *L* and *L′* ever intersect?

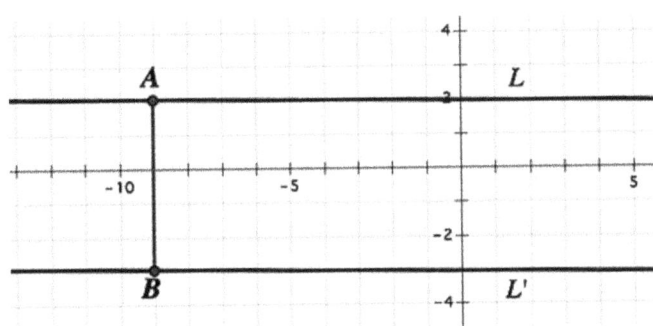

Ⓐ Yes, line *L′* is now the same as *L*.
Ⓑ Yes, parallel lines always eventually intersect.
Ⓒ No, every point on *L′* will always have a corresponding point the distance of $\overline{AB}$ away from *L′*.
Ⓓ No, the translation along $\overline{AB}$ does not change the slope from *L* to *L′*, and lines with the same slope never intersect.

5. Line *L* is translated along segment $\overline{AB}$ to create line *L′*. Will *L* and *L′* ever intersect?

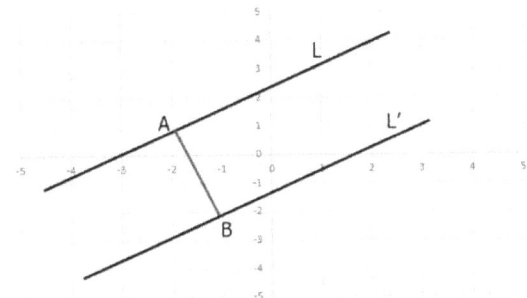

Ⓐ Yes, line *L′* is now the same as *L*.
Ⓑ Yes, parallel lines always eventually intersect.
Ⓒ Yes, every point on *L′* will not always have a corresponding point the distance of AB away from *L′*.
Ⓓ No, the translation along the line segment *AB* does not change the slope from *L* to *L′*, and lines with the same slope never intersect.

6. Line *L* is translated along ray *AC* to create line *L'*. What do you know about the relationship between line *L* and *L'*?

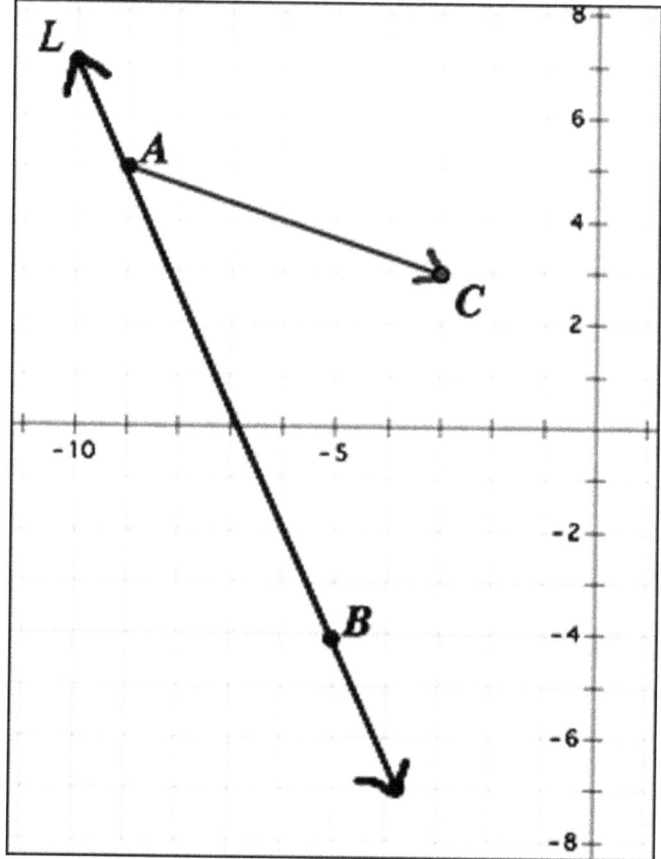

   Ⓐ The lines intersect at least once.
   Ⓑ The lines are exactly the same.
   Ⓒ The lines are parallel.
   Ⓓ None of the above.

LumosLearning.com

7. **How many lines can be drawn through point *C* that are parallel to line *L*?**

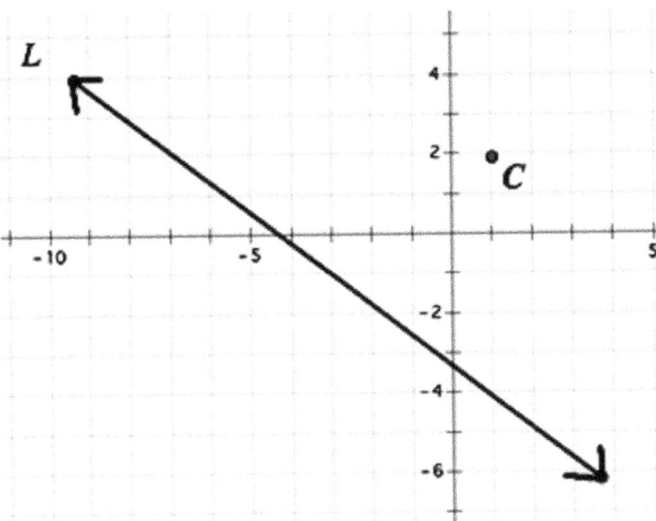

Ⓐ None
Ⓑ One
Ⓒ Two
Ⓓ Infinitely many

8. **Figure *ABCD* was rotated around the origin to create *WXYZ*. Prove *XY* and *WZ* are parallel.**

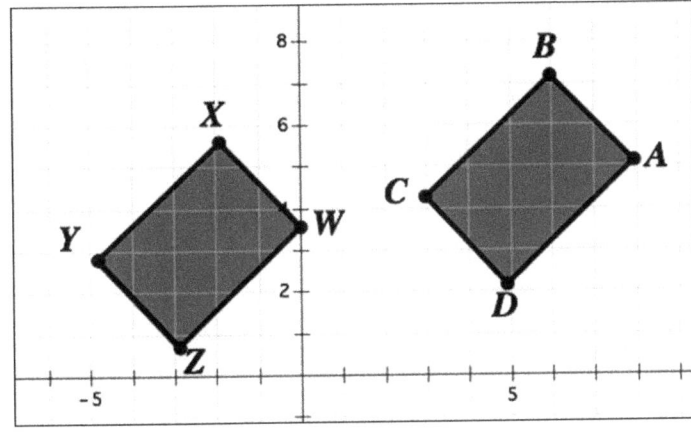

Ⓐ The sides of all rectangles are parallel.

Ⓑ Since *ABCD* is a rectangle, *AD* ||*BC*. Translations map parallel lines to parallel lines, so *XY* || *WZ*.

Ⓒ It cannot be proven because *XY* and *WZ* are perpendicular.

Ⓓ It cannot be proven because the angle of rotation is not given.

LumosLearning.com

9. The two parallel lines shown are rotated 180° abound the origin. What is the result of this transformation?

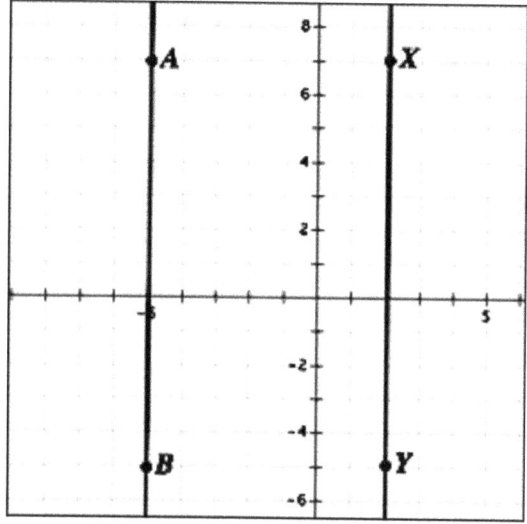

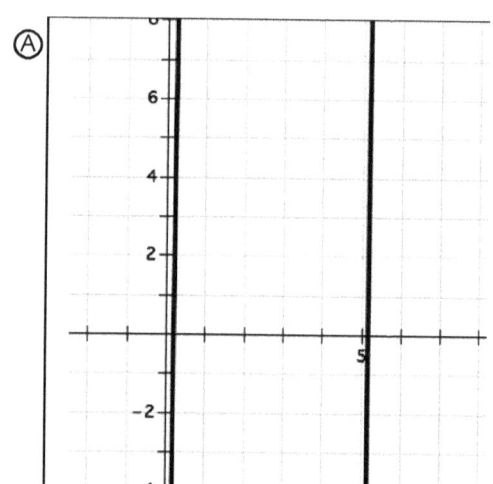

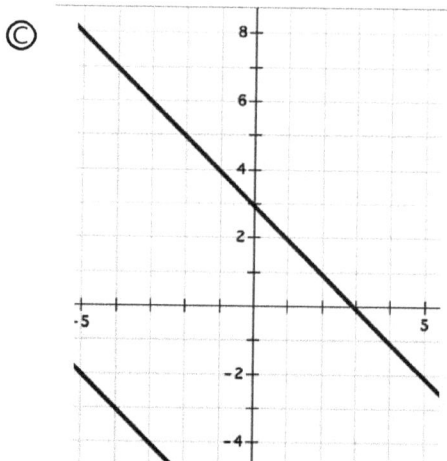

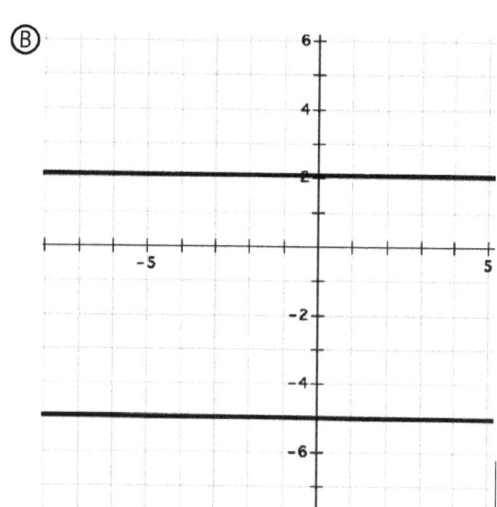

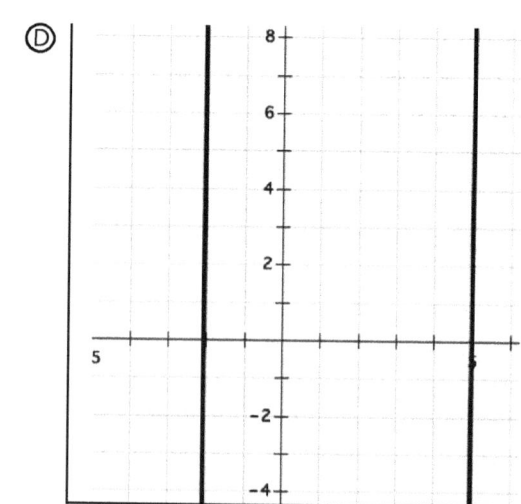

LumosLearning.com

10. Figure *DEF* is the result of 180° rotation around the origin of figure *ABC*. Prove $\overline{AB}$ and $\overline{DE}$ are parallel.

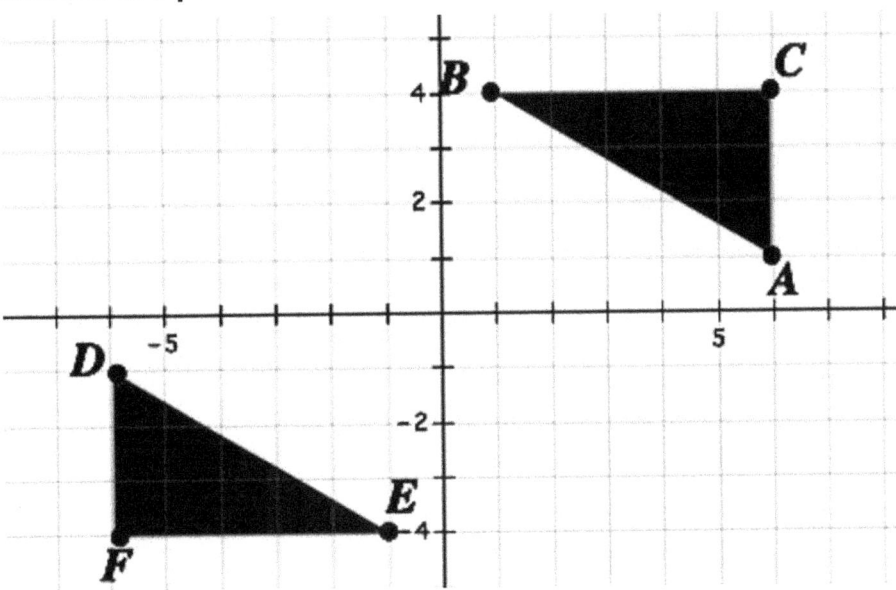

Ⓐ A 180° rotation of a given segment always maps to a parallel segment. Therefore $\overline{AB} \parallel \overline{DE}$.

Ⓑ Corresponding sides of triangles are always parallel. Therefore, $\overline{AB} \parallel \overline{DE}$.

Ⓒ It cannot be proven because $\overline{AB}$ and $\overline{DE}$ are perpendicular.

Ⓓ It cannot be proven due to the angle of rotation is not given.

11. Select what concepts are preserved under these different transformations. Select all that apply.

	Lengths of sides	Angle Measures	Parallel Sides on Figure
Translation	○	○	○
Reflection	○	○	○
Rotation	○	○	○
Dilation	○	○	○

12. Figure ABCD undergoes the shown transformation. The slope of $\overline{AC}$ is $-\frac{1}{3}$. What is the slope of A'C'?

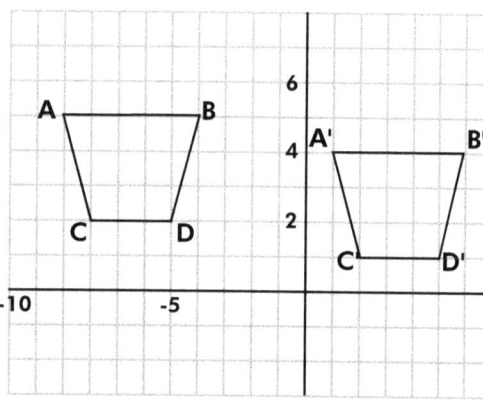

Slope of A'C' is ☐

13. Select the one that correctly shows the parallel lines that have been correctly reflected over the y-axis.
    Circle the correct answer choice

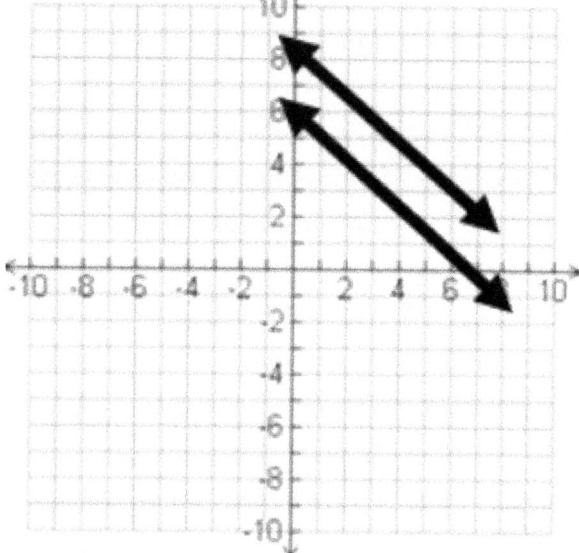

Ⓐ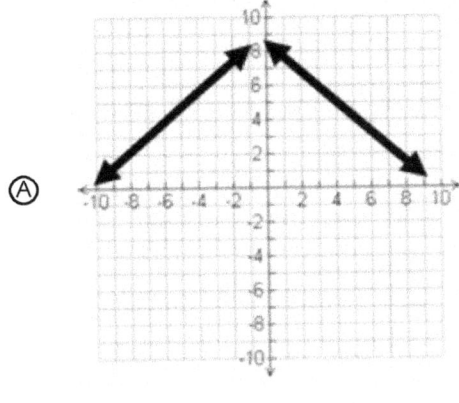

Ⓑ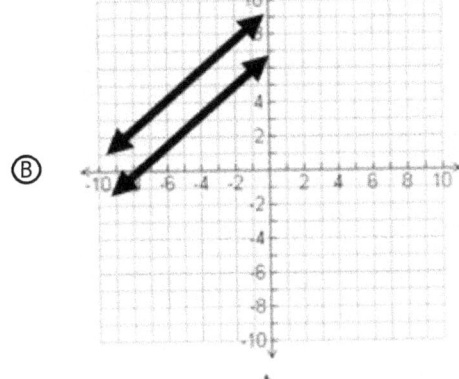

Ⓒ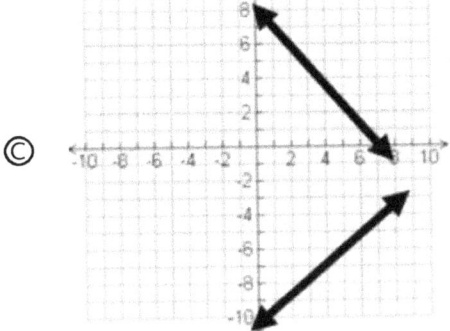

Did You Check Your Score?

YES	NO
Record your score below: Score (%): _____ Date: _____	► Scan the QR code or Visit *lumoslearning.com/a/8m021* ► Submit your answers using the *Online Answer Sheet.* ► Get your Scores & Detailed Explanations.

DIRECTIONS

Do **NOT** write your answers in this book. **OPEN** the Online Answer Sheet by Scanning the **QR Code** or Visit **lumoslearning.com/a/8m022**

Chapter 4 → Lesson 4: Transformations of Congruency

1. Which of the following examples best represents congruency in nature?

   Ⓐ A mother bear and her cub.
   Ⓑ The wings of a butterfly.
   Ⓒ The tomatoes were picked from my garden.
   Ⓓ The clouds in the sky.

2. If triangle ABC is drawn on a coordinate plane and then reflected over the vertical axis, which of the following statements is true?

   Ⓐ The reflected triangle will be similar ONLY to the original.
   Ⓑ The reflected triangle will be congruent to the original.
   Ⓒ The reflected triangle will be larger than the original.
   Ⓓ The reflected triangle will be smaller than the original.

3. What transformation was applied to the object in quadrant 2 to render the results in the graph below?

   Ⓐ reflection
   Ⓑ rotation
   Ⓒ translation
   Ⓓ not enough information

LumosLearning.com

4. What transformation was applied to the object in quadrant 2 to render the results in the graph below?

Ⓐ reflection
Ⓑ rotation
Ⓒ translation
Ⓓ not enough information

5. What is NOT true about the graph below?

Ⓐ The object in quadrant 3 could not be a reflection of the object in quadrant 2.
Ⓑ The object in quadrant 3 could be a translation of the object in quadrant 2.
Ⓒ The two objects are congruent.
Ⓓ The object in quadrant 3 could not be a dilation of the object in quadrant 2.

6. Finish the statement. Two congruent objects _____.

Ⓐ have the same dimensions.
Ⓑ have different measured angles.
Ⓒ are not the same shape.
Ⓓ only apply to two-dimensional objects.

7. What transformations can be applied to an object to create a congruent object?

Ⓐ all transformations
Ⓑ dilation and rotation
Ⓒ translation and dilation
Ⓓ reflection, translation, and rotation

8. A figure formed by rotation followed by reflection of an original triangle will be _____.

Ⓐ similar only and not congruent to the original.
Ⓑ congruent to the original.
Ⓒ smaller than the original.
Ⓓ larger than the original.

9. Which of the following is NOT a characteristic of congruent triangles?

Ⓐ They have three pairs of congruent sides.
Ⓑ They have three pairs of congruent angles.
Ⓒ Their areas are equal.
Ⓓ They have four pairs of proportional sides

10. Which of the following letters looks the same after a reflection followed by a 180° rotation.

Ⓐ P
Ⓑ O
Ⓒ F
Ⓓ None of the above.

11. Select all that apply to this transformation.

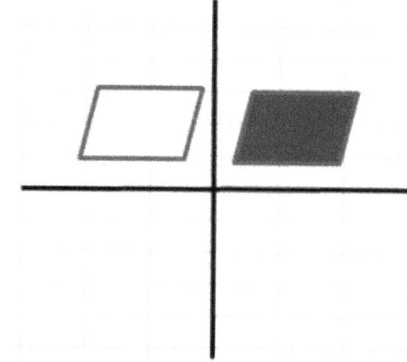

	Apply	Doesn't Apply
the two shapes are congruent	○	○
the two shapes are not congruent	○	○
the two shapes are similar	○	○
the two shapes have the same size	○	○
one shape is rotated from the other shape	○	○
one shape is reflected from the other shape	○	○
one shape is translated from the other shape	○	○

12. Quadrilateral ABCD is translated 5 units to the left and 4 units down. Which congruent quadrilateral match this transformation?

Write your answer in the box given below

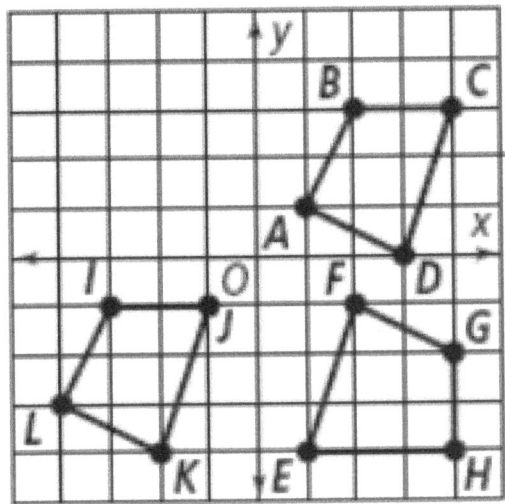

```
┌──────────────────────────────────────────┐
│                                            │
│                                            │
│                                            │
│                                            │
└──────────────────────────────────────────┘
```

Did You Check Your Score?

YES	NO
Record your score below:	► Scan the QR code or Visit *lumoslearning.com/a/8m022*
Score (%): _____	► Submit your answers using the *Online Answer Sheet*.
Date: _____	► Get your Scores & Detailed Explanations.

DIRECTIONS

Do **NOT** write your answers in this book. **OPEN** the Online Answer Sheet by Scanning the **QR Code** or Visit **lumoslearning.com/a/8m023**

Chapter 4 → Lesson 5: Analyzing Transformations

1. **Which of the following transformations could transform triangle A to triangle B?**

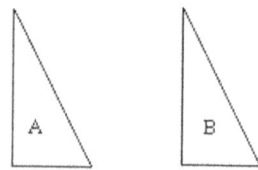

Ⓐ **Rotation**
Ⓑ **Reflection**
Ⓒ **Translation**
Ⓓ **Dilation**

2. **Which of the following transformations could transform triangle A to triangle B?**

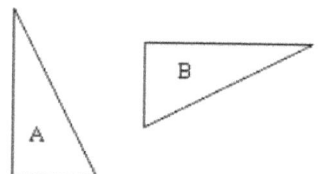

Ⓐ **Rotation**
Ⓑ **Reflection**
Ⓒ **Translation**
Ⓓ **Dilation**

3. **Which of the following transformations could transform triangle A to triangle B?**

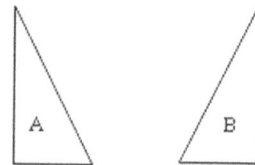

Ⓐ **Rotation**
Ⓑ **Reflection**
Ⓒ **Translation**
Ⓓ **Dilation**

LumosLearning.com

4. Which of the following transformations could transform triangle A to triangle B?

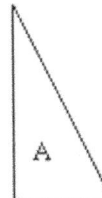 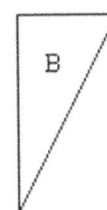

    Ⓐ Rotation
    Ⓑ Reflection
    Ⓒ Translation
    Ⓓ Combination of above

5. Which of the following transformations could transform triangle A to triangle B?

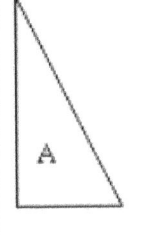 

    Ⓐ Rotation
    Ⓑ Reflection
    Ⓒ Translation
    Ⓓ Dilation

6. Which of the following transformations could transform triangle A to triangle B?

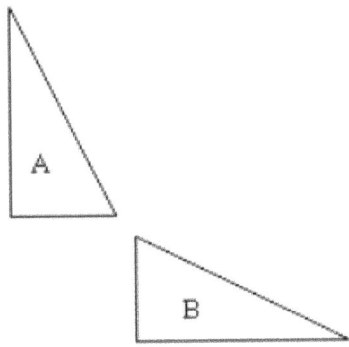

    Ⓐ Rotation
    Ⓑ Reflection
    Ⓒ Translation
    Ⓓ Dilation

7. Which of the following sequences of transformations could transform triangle A to triangle B?

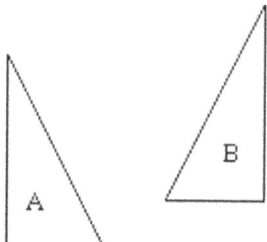

- Ⓐ A reflection followed by a translation
- Ⓑ A reflection followed by another reflection
- Ⓒ A rotation followed by a translation
- Ⓓ A dilation followed by a translation

8. Which of the following transformations does NOT preserve congruency?

- Ⓐ Rotation
- Ⓑ Translation
- Ⓒ Reflection
- Ⓓ Dilation

9. Consider the triangle with vertices (1, 0), (2, 5) and (-1, 5). Find the vertices of the new triangle after a reflection over the vertical axis followed by a reflection over the horizontal axis.

- Ⓐ (-1, 0), (1, 5) and (-2, 5)
- Ⓑ (-1, 0), (-2, -5) and (1, -5)
- Ⓒ (1, 0), (-2, 5) and (1, 5)
- Ⓓ (1, 0), (-2, -5) and (1, -5)

10. Translate the triangle with vertices (1, 0), (2, 5), and (-1, 5), 3 units to the left. Which of the following ordered pairs represent the vertices of the new triangle?

- Ⓐ (-2, 0), (-4, 5) and (-1, 5)
- Ⓑ (4, 0), (5, 5) and (2, 5)
- Ⓒ (-1, 2), (2, 2) and (1, -3)
- Ⓓ (-1, 8), (2, 8) and (1, 3)

11. Select the coordinates that will correspond with each transformation for point A.

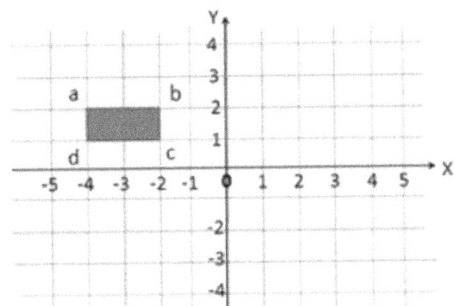

	A(4,-2)	A(-2,1)	A(-4,-2)
Translation (x+2,y-1)	○	○	○
Rotation 180°	○	○	○
Reflection over x-axis	○	○	○

12. In the coordinate plane shown, ΔABC has vertices A(7, 6), B(4, 2), and C(10, 2). What scale factor was used on ΔABC to get ΔDEF.

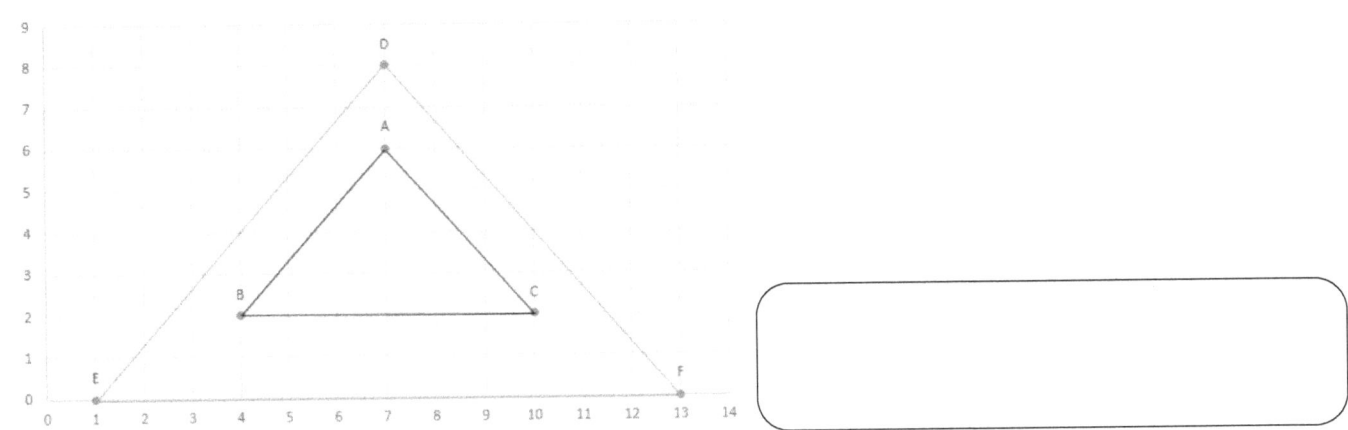

Did You Check Your Score?

YES	NO
Record your score below: Score (%): _____ Date: _____	► Scan the QR code or Visit *lumoslearning.com/a/8m023* ► Submit your answers using the *Online Answer Sheet*. ► Get your Scores & Detailed Explanations.

DIRECTIONS

Do **NOT** write your answers in this book. **OPEN** the Online Answer Sheet by Scanning the **QR Code** or Visit **lumoslearning.com/a/8m024**

Chapter 4 → Lesson 6: Transformations & Similarity

1. Which of the following transformations could transform triangle A to triangle B?

- Ⓐ Rotation
- Ⓑ Reflection
- Ⓒ Translation
- Ⓓ Dilation

2. What transformations have been applied to the large object to render the results in the graph below?

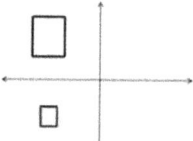

- Ⓐ reflection and dilation
- Ⓑ rotation and translation
- Ⓒ rotation and dilation
- Ⓓ translation and reflection

3. What transformations have been applied to the large object to render the results in the graph below?

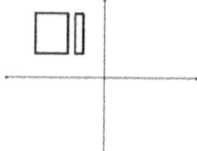

- Ⓐ no transformation
- Ⓑ reflection and dilation
- Ⓒ translation and dilation
- Ⓓ rotation and dilation

LumosLearning.com

4. Which graph represents reflection over an axis and dilation?

Ⓐ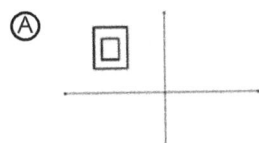

Ⓑ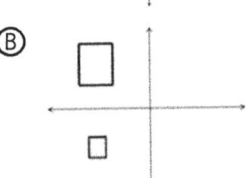

Ⓒ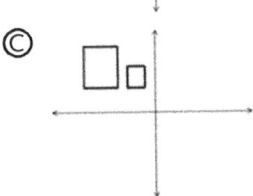

Ⓓ None of the above.

5. What transformation is necessary to have two similar, but not congruent, objects?

Ⓐ Rotation
Ⓑ Translation
Ⓒ Dilation
Ⓓ Reflection

6. Finish this statement. Two similar objects _____.

Ⓐ have proportional dimensions.
Ⓑ are always congruent.
Ⓒ have different measured angles.
Ⓓ can be different shapes.

7. If a point, P, on a coordinate plane moves from (9, 3) to P' (3, -9) and then to P'' (0, -9), what transformations have been applied?

Ⓐ Dilation followed by Translation
Ⓑ Translation followed by Dilation
Ⓒ Rotation followed by Translation
Ⓓ Translation followed by Rotation

8. Consider Triangle ABC, where AB = 5, BC = 3, and AC = 6, and Triangle WXY, where WX = 10, XY = 6, and WY = 12.
Assume ABC is similar to WXY.
Which of the following represents the ratio of similarity?

Ⓐ 1 : 2
Ⓑ 5 : 6
Ⓒ 3 : 10
Ⓓ 6 : 20

9. Rectangle A is 1 unit by 2 units.
Rectangle B is 2 units by 3 units.
Rectangle C is 2 units by 4 units.
Rectangle D is 3 units by 6 units.
Which rectangle is not similar to the other three rectangles?

Ⓐ A
Ⓑ B
Ⓒ C
Ⓓ D

10. If triangle ABC is similar to triangle WXY and AB = 9, BC = 7, AC = 14, WX = 27, and XY = 21.
Find WY.

Ⓐ 44
Ⓑ 43
Ⓒ 42
Ⓓ 41

11. Select which transformations were used to map the pre-image onto the image. Also select if the transformation used leaves the figure congruent or if it only makes them similar.

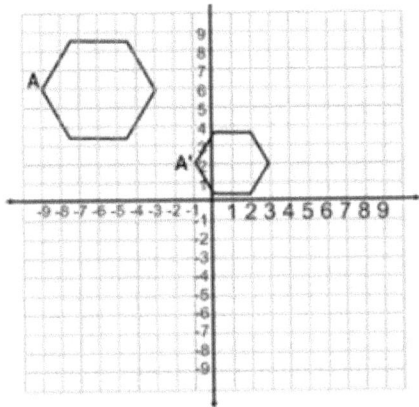

	Used	Similar only	Congruent
Translation	○	○	○
Rotation	○	○	○
Reflection	○	○	○
Dilation	○	○	○

12. The 2 figures are similar. What is the height of the 2nd figure? Write your answer in the box below.

Did You Check Your Score?

YES	NO
Record your score below: Score (%): _____ Date: _____	► Scan the QR code or Visit *lumoslearning.com/a/8m024* ► Submit your answers using the *Online Answer Sheet.* ► Get your Scores & Detailed Explanations.

DIRECTIONS

Do **NOT** write your answers in this book. **OPEN** the Online Answer Sheet by Scanning the **QR Code** or Visit **lumoslearning.com/a/8m025**

Chapter 4 → Lesson 7: Interior & Exterior Angles in Geometric Figures

1. What term describes a pair of angles formed by the intersection of two straight lines that share a common vertex but do not share any common sides?

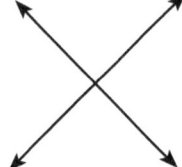

   Ⓐ Supplementary Angles
   Ⓑ Complementary Angles
   Ⓒ Horizontal Angles
   Ⓓ Vertical Angles

2. If a triangle has two angles with measures that add up to 100 degrees, what must the measure of the third angle be?

   Ⓐ 180 degrees
   Ⓑ 100 degrees
   Ⓒ 80 degrees
   Ⓓ 45 degrees

3.

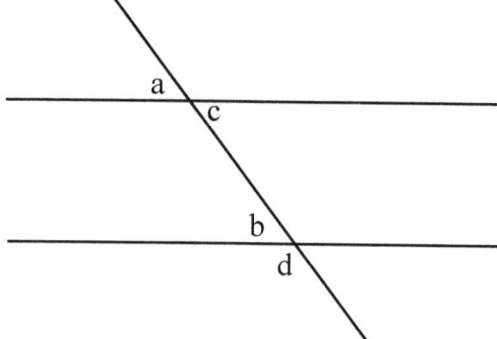

   The figure shows two parallel lines intersected by a third line. If a = 55°, what is the value of b?

   Ⓐ 35°
   Ⓑ 45°
   Ⓒ 55°
   Ⓓ 125°

4.

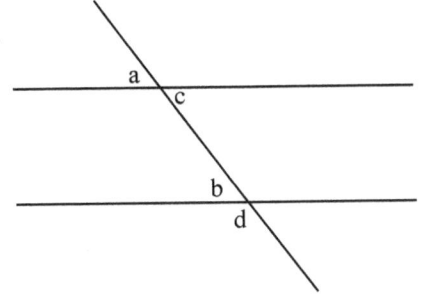

The figure shows two parallel lines intersected by a third line. If b = 60°, what is the value of c?

Ⓐ 30°
Ⓑ 60°
Ⓒ 90°
Ⓓ 120°

5.

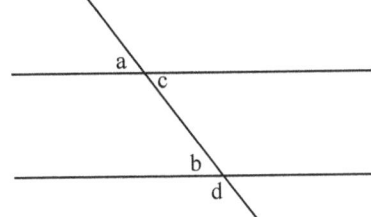

The figure shows two parallel lines intersected by a third line. If d = 130°, what is the value of a?

Ⓐ 30°
Ⓑ 40°
Ⓒ 50°
Ⓓ 130°

6.

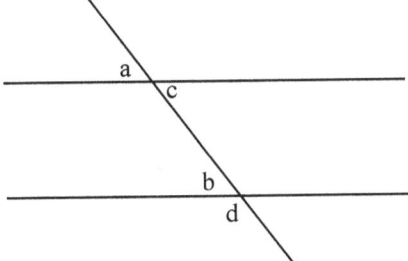

The figure shows two parallel lines intersected by a third line. Which of the following angles are equal in measure?

Ⓐ a and b only
Ⓑ a and c only
Ⓒ b and c only
Ⓓ a, b, and c

7. Two angles in a triangle measure 65° each. What is the measure of the third angle in the triangle?

Ⓐ 25°
Ⓑ 50°
Ⓒ 65°
Ⓓ 130°

8.

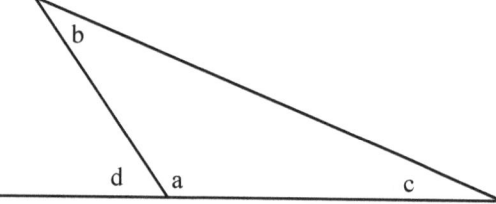

If b = 40° and c = 30°, what is the measure of d?

Ⓐ 35°
Ⓑ 70°
Ⓒ 110°
Ⓓ 145°

9. In right triangle ABC, Angle C is the right angle. Angle A measures 70°. Find the measure of the exterior angle at angle C.

Ⓐ 180°
Ⓑ 90°
Ⓒ 110°
Ⓓ 160°

10. If two parallel lines are cut by a transversal, the alternate interior angles are _____.

Ⓐ supplementary
Ⓑ complementary
Ⓒ equal in measure
Ⓓ none of the above

LumosLearning.com

11. Match the figure with the sum of the interior angles of each polygon.

	2520	1080	1440	4140	540
Decagon	○	○	○	○	○
16-gon	○	○	○	○	○
Pentagon	○	○	○	○	○
25-gon	○	○	○	○	○
Octagon	○	○	○	○	○

12. Observe the figure given below. ∠3 and ∠7 are what type of angles?

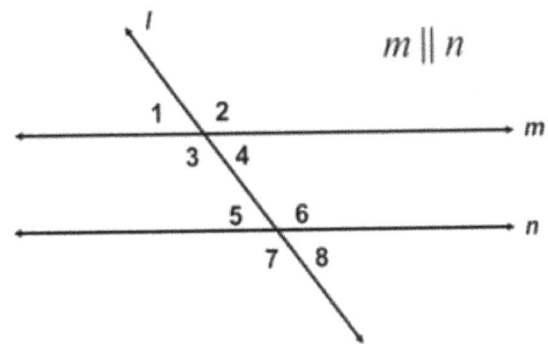

∠3 and ∠7 are what type of angles? Write your answer in the box given below.

Did You Check Your Score?

YES	NO
Record your score below: Score (%): _____ Date: _____	► Scan the QR code or Visit *lumoslearning.com/a/8m025* ► Submit your answers using the *Online Answer Sheet*. ► Get your Scores & Detailed Explanations.

LumosLearning.com

DIRECTIONS

Do **NOT** write your answers in this book. **OPEN** the Online Answer Sheet by Scanning the **QR Code** or Visit **lumoslearning.com/a/8m026**

Chapter 4 → Lesson 8: Verifying the Pythagorean Theorem

1. Which of the following could be the lengths of the sides of a right triangle?

   Ⓐ 1, 2, 3
   Ⓑ 2, 3, 4
   Ⓒ 3, 4, 5
   Ⓓ 4, 5, 6

2. A triangle has sides 8 cm long and 15 cm long, with a 90° angle between them. What is the length of the third side?

   Ⓐ 7 cm
   Ⓑ 17 cm
   Ⓒ 23 cm
   Ⓓ 289 cm

3. Find the value of c, rounded to the nearest tenth.

   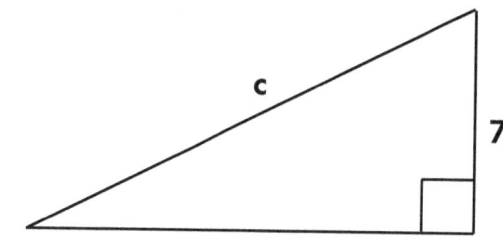

   Ⓐ 8.5
   Ⓑ 8.8
   Ⓒ 13.0
   Ⓓ 19.3

4. A square has sides 5 inches long. What is the approximate length of a diagonal of the square?

   Ⓐ 5 inches
   Ⓑ 6 inches
   Ⓒ 7 inches
   Ⓓ 8 inches

5. Which of the following INCORRECTLY completes this statement of the Pythagorean theorem?
    In a right triangle with legs of lengths a and b and hypotenuse of length c, ...

    Ⓐ $a^2 + b^2 = c^2$
    Ⓑ $c^2 - a^2 = b^2$
    Ⓒ $c^2 - b^2 = a^2$
    Ⓓ $a^2 + c^2 = b^2$

6. A Pythagorean triplet is a set of three positive integers a, b, and c that satisfy the equation $a^2 + b^2 = c^2$. Which of the following is a Pythagorean triple?

    Ⓐ   a = 3, b = 6, c = 9
    Ⓑ   a = 6, b = 9, c = 12
    Ⓒ   a = 9, b = 12, c = 15
    Ⓓ   a = 12, b = 15, c = 18

7. If an isosceles right triangle has legs of 4 inches each, find the length of the hypotenuse.

    Ⓐ   Approximately 6 in.
    Ⓑ   Approximately 5 in.
    Ⓒ   Approximately 4 in.
    Ⓓ   Approximately 3 in.

8. In triangle ABC, angle C = 90°, AC = 4 and AB = 10. Find BC to the nearest tenth.

    Ⓐ   9.5
    Ⓑ   9.2
    Ⓒ   8.9
    Ⓓ   8.5

9. In triangle ABC, angle C = 90°, AB = 35, and BC = 28. Find AC.

    Ⓐ 23
    Ⓑ 22
    Ⓒ 21
    Ⓓ 20

10. The diagonal of a square is 25. Find the approximate side lengths.

    Ⓐ 15
    Ⓑ 16
    Ⓒ 17
    Ⓓ 18

LumosLearning.com

11. You have a right triangle with leg lengths of 6 and 8. What is the length of the hypotenuse? Fill in the numbers into the equation and solve.

$6^2 + 8^2 = C^2$

$100 = C^2$

$\sqrt{100} = C$

$C = $ _____

12. Which equation would you use to solve for the missing side of the triangle pictured below?

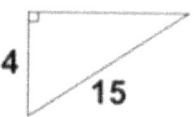

Ⓐ $4^2 + 15^2 = x^2$

Ⓑ $4^2 + x^2 = 15^2$

Ⓒ $x^2 + 15^2 = 4^2$

 **Did You Check Your Score?**

YES	NO
Record your score below:	► Scan the QR code or Visit *lumoslearning.com/a/8m026*
Score (%): _____	► Submit your answers using the *Online Answer Sheet*.
Date: _____	► Get your Scores & Detailed Explanations.

LumosLearning.com

DIRECTIONS

Do **NOT** write your answers in this book. **OPEN** the Online Answer Sheet by Scanning the **QR Code** or Visit **lumoslearning.com/a/8m027**

Chapter 4 → Lesson 9: Pythagorean Theorem in Real-World Problems

1. The bottom of a 17-foot ladder is placed on level ground 8 feet from the side of a house as shown in the figure below. Find the vertical height at which the top of the ladder touches the side of the house.

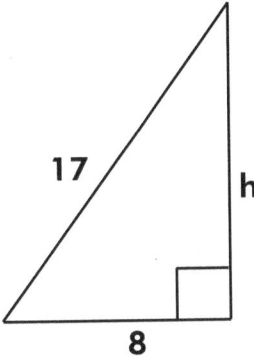

Ⓐ h = 9 feet
Ⓑ h = 12 feet
Ⓒ h = 15 feet
Ⓓ h = 18 feet

2. Which of the following equations could be used to find the value of w?

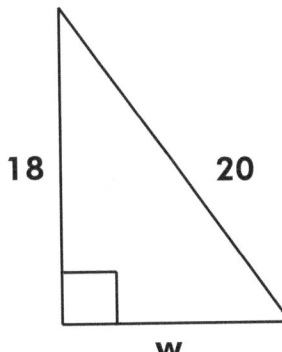

Ⓐ $w^2 + 18^2 = 20^2$
Ⓑ $18^2 - w^2 = 20^2$
Ⓒ $20^2 + 18^2 = w^2$
Ⓓ $w + 18 = 20$

3. John has a chest where he keeps his antiques. What is the measure of the diagonal (d) of John's chest with the height (c) = 3ft, width (b) = 3ft, and length (a) = 5ft.?

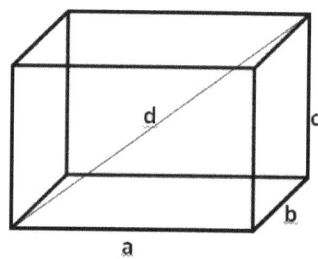

Ⓐ √42 ft²
Ⓑ √44 ft²
Ⓒ √34 ft
Ⓓ √43 ft

4. Mary has a lawn that has a width (a) of 30 yards, and a length (b) of 40 yards. What is the measurement of the diagonal (c) of the lawn?

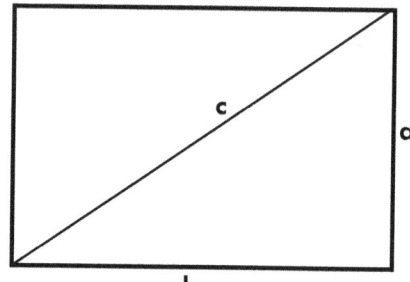

Ⓐ 50 yards
Ⓑ 49 yards
Ⓒ 51 yards
Ⓓ 50 yards²

5. A construction company needed to build a sign with the width (a) of 9 ft, and a length (b) of 20 ft. What will be the approximate measurement of the diagonal (c) of the sign?

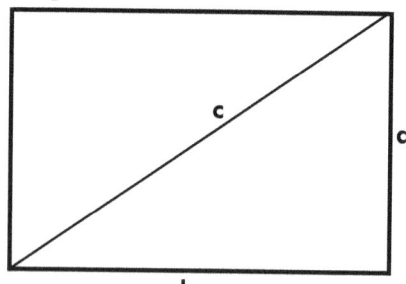

Ⓐ 23 ft
Ⓑ 22 ft
Ⓒ 20 ft
Ⓓ 19 ft

LumosLearning.com

6. An unofficial baseball diamond is measured to be 50 yards wide. What is the approximate measurement of one side (a) of the diamond?

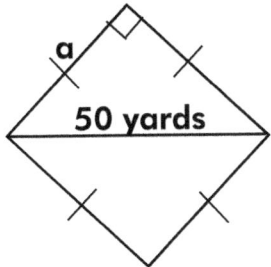

   Ⓐ 34 ft
   Ⓑ 35 yards
   Ⓒ 35 ft
   Ⓓ 36 yards

7. The neighborhood swimming pool is 20 ft wide and 30 ft long. What is the approximate measurement of the diagonal (d) of the base of the pool?

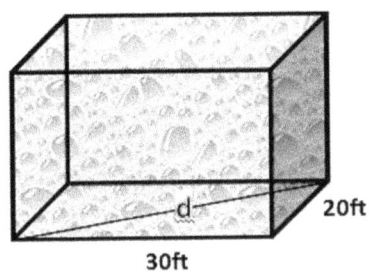

   Ⓐ 36 ft
   Ⓑ 35 ft
   Ⓒ 34 ft
   Ⓓ 34 yards

8. To get to her friend's house, a student must walk 20 feet to the corner of their streets, turn left and walk 15 feet to her friend's house.
    How much shorter would it be if she could cut across a neighbor's yard and walk a straight line from her house to her friend's house?

   Ⓐ 5 feet shorter
   Ⓑ 10 feet shorter
   Ⓒ 25 feet shorter
   Ⓓ 35 feet shorter

9. Your school principal wants the custodian to put a new flag up on the flagpole. If the flagpole is 40 feet tall and they have a 50 foot ladder, approximately how far from the base of the pole can he place the base of his ladder in order to accomplish the task?

Ⓐ Up to 10 ft away
Ⓑ Up to 20 ft away
Ⓒ Up to 30 ft away
Ⓓ Up to 40 ft away

10. Your kite is at the end of a 50 ft string. You are 25 ft from the outside wall of a building that you know makes a right angle with the ground.
How high is the kite approximately?

Ⓐ Approximately 39 ft high
Ⓑ Approximately 40 ft high
Ⓒ Approximately 42 ft high
Ⓓ Approximately 43 ft high

11. Match the following word problem with the correct equation that you would use to solve it.

	$9^2+x^2=15^2$	$40^2+38^2=x^2$	$9^2+15^2=x^2$	$x^2+38^2=40^2$
One house is 15 miles due north of the park. Another house is 9 miles due east of the park. How far apart are the houses from each other?	○	○	○	○
The foot of a ladder is put 9 feet from the wall. If the ladder is 15 feet long how high up the building will the ladder reach?	○	○	○	○
If you drive your car 40 miles south and then 38 miles east, how far would the shortest route be from your starting point?	○	○	○	○
The diagonal of a TV is 40 inches. The TV is 38 inches long. How tall is the TV?	○	○	○	○

12. Fill in the missing information needed to solve for the word problem. Round to the nearest tenth if necessary.

WORD PROBLEM	LEG(a)	LEG(b)	HYPOTENUSE(c)
Find the height of a pyramid whose slant height is 26 cm and base length is 48 cm	24		
Find the base length of a pyramid whose height is 8 in and slant height 17 in.		8	17
The foot of a ladder is put 5 feet from the wall. If the top of the ladder is 10 feet from the ground, how long is the ladder?		10	

13. On a bike ride you start at your house and ride your bike 3.5 miles north and then 1.25 miles east. How far are you directly from your house? Select the correct equation with the correct solution to match the word problem.

Circle the correct answer choice.

Ⓐ $\sqrt{3.5^2 + 1.25^2} = 22.56$ mi
Ⓑ $\sqrt{3.5^2 - 1.25^2} = 3.34$ mi
Ⓒ $\sqrt{3.5^2 + 1.25^2} = 3.72$ mi

Did You Check Your Score?

YES	NO
Record your score below: Score (%): _____ Date: _____	► Scan the QR code or Visit *lumoslearning.com/a/8m027* ► Submit your answers using the *Online Answer Sheet*. ► Get your Scores & Detailed Explanations.

DIRECTIONS

Do **NOT** write your answers in this book. **OPEN** the Online Answer Sheet by Scanning the **QR Code** or Visit **lumoslearning.com/a/8m028**

Chapter 4 → Lesson 10: Pythagorean Theorem & Coordinate System

1. A robot begins at point A, travels 4 meters west, then turns and travels 7 meters south, reaching point B. What is the approximate straight-line distance between points A and B?

   Ⓐ 8 meters
   Ⓑ 9 meters
   Ⓒ 10 meters
   Ⓓ 11 meters

2. What is the distance between the points (1, 3) and (9, 9)?

   Ⓐ 6 units
   Ⓑ 8 units
   Ⓒ 10 units
   Ⓓ 12 units

3. Find the distance (approximately) between Pt A (2, 7) and Pt B (-2, -7).

   Ⓐ 14.0
   Ⓑ 14.6
   Ⓒ 18.0
   Ⓓ 13.4

4. Find the distance (approximately) between Pt P (5, 3) and the origin (0, 0).

   Ⓐ 4.0
   Ⓑ 5.1
   Ⓒ 5.8
   Ⓓ 8.0

5. Find the distance (approximately) between the points A (11, 12) and B (7, 8).

   Ⓐ 4.0
   Ⓑ 5.3
   Ⓒ 5.7
   Ⓓ 8.0

6. Is it closer to go from Pt A (4, 6) to Pt B (2, -4) or Pt A to Pt C (-5, 2)?

   Ⓐ A to B
   Ⓑ A to C
   Ⓒ Neither, they are both the same distance.
   Ⓓ Not enough information.

7. Paul lives 50 yards east and 40 yards south of his friend, Larry. If he wants to shorten his walk by walking in a straight line from his home to Larry's, how far (approximately) will he walk?

   Ⓐ 64 yards
   Ⓑ 62 yards
   Ⓒ 60 yards
   Ⓓ 58 yards

8. In a coordinate plane, a point P (7,8) is rotated 90° clockwise around the origin and then reflected across the vertical axis. Find the distance (approximately) between the original point and the final point.

   Ⓐ 14 units
   Ⓑ 21 units
   Ⓒ 24 units
   Ⓓ None of the above

9. You are using a coordinate plane to sketch out a plan for your vegetable garden. Your garden will be a rectangle 15 ft wide and 20 ft long. You want a square in the center with 3 ft sides to be reserved for flowers. If the garden is plotted on the coordinate plane with the southwest corner at the origin, what are the coordinates of the center of the flower garden?

   Ⓐ (15, 10)
   Ⓑ (7.5, 10)
   Ⓒ (-15, 10)
   Ⓓ (-7.5, 10)

LumosLearning.com

10. You are using a coordinate plane to sketch out a plan for a vegetable garden. The garden will be a rectangle 15 feet wide and 20 feet long. You want a square in the center with 3 feet sides to be reserved for flowers. The garden is plotted in the coordinate grid so that the southwest corner is placed at the origin. Find the length of the diagonal (approximately) of the flower bed.

Ⓐ 4.0 feet
Ⓑ 4.2 feet
Ⓒ 5.0 feet
Ⓓ 5.2 feet

11. Match the ordered pairs with the approximate distance between them.

	10.8	12.2	13	14.8
(6, 5) and (-4, 9)	○	○	○	○
(-8, 0) and (5, -7)	○	○	○	○
(-4, -9) and (6, -2)	○	○	○	○
(5, 4) and (12, 15)	○	○	○	○

12. Fill in the missing values. Round to the nearest hundredth if necessary.

ORDERED PAIRS	LEG LENGTH	LEG LENGTH	HYPOTENUSE
(6,2),(0,-6)	6		
(-3,-1),(-4,0)	1	1	
(-2,3),(-1,7)			4.12

Did You Check Your Score?

YES	NO
Record your score below: Score (%): _____ Date: _____	► Scan the QR code or Visit *lumoslearning.com/a/8m028* ► Submit your answers using the *Online Answer Sheet*. ► Get your Scores & Detailed Explanations.

DIRECTIONS

Do **NOT** write your answers in this book. **OPEN** the Online Answer Sheet by Scanning the **QR Code** or Visit **lumoslearning.com/a/8m029**

Chapter 4 → Lesson 11: Finding Volume: Cone, Cylinder, & Sphere

1. **What is the volume of a sphere with a radius of 6?**

   Ⓐ 72π
   Ⓑ 144π
   Ⓒ 216π
   Ⓓ 288π

2. **A cone has a height of 9 and a base whose radius is 4. Find the volume of the cone.**

   Ⓐ 18π
   Ⓑ 36π
   Ⓒ 48π
   Ⓓ 72π

3. **What is the volume of a cylinder with a radius of 5 and a height of 3?**

   Ⓐ 30π
   Ⓑ 45π
   Ⓒ 75π
   Ⓓ 120π

4. **A round ball has a diameter of 10 inches. What is the approximate volume of the ball? Use $\pi = 3.14$**

   Ⓐ 130 cubic inches
   Ⓑ 260 cubic inches
   Ⓒ 390 cubic inches
   Ⓓ 520 cubic inches

5. A cylindrical can has a height of 5 inches and a diameter of 4 inches. What is the approximate volume of the can? Use $\pi = 3.14$

   Ⓐ 20 cubic inches
   Ⓑ 60 cubic inches
   Ⓒ 100 cubic inches
   Ⓓ 200 cubic inches

6. A cylinder and a cone have the same radius and the same volume. How do the heights compare?

   Ⓐ The height of the cylinder is 3 times the height of the cone.
   Ⓑ The height of the cylinder is 2 times the height of the cone.
   Ⓒ The height of the cone is 2 times the height of the cylinder.
   Ⓓ The height of the cone is 3 times the height of the cylinder.

7. Which of the following has the greatest volume?

   Ⓐ A sphere with a radius of 2 cm
   Ⓑ A cylinder with a height of 2 cm and a radius of 2 cm
   Ⓒ A cone with a height of 4 cm and a radius of 3 cm
   Ⓓ All three volumes are equal

8. Which of the following has the greatest volume?

   Ⓐ A sphere with a radius of 3 cm
   Ⓑ A cylinder with a height of 4 cm and a radius of 3 cm
   Ⓒ A cone with a height of 3 cm and a radius of 6 cm
   Ⓓ All three volumes are equal

9. If a sphere and a cone have the same radii r and the cone has a height of 4, find the ratio of the volume of the sphere to the volume of the cone.

   Ⓐ r : 1
   Ⓑ 1 : r
   Ⓒ r : 4
   Ⓓ 4 : r

10. Which of the following describes the relationship between the volumes of a cone and cylinder with the same radii and the same height.

   Ⓐ The volume of the cylinder is three times that of the cone.
   Ⓑ The volume of the cylinder is 1/3 that of the cone.
   Ⓒ The volume of the cylinder is 4/3 that of the cone.
   Ⓓ None of the above.

11. **Find the volume of the figure below. Use pi = 3.14. Write your answer in the box below.**

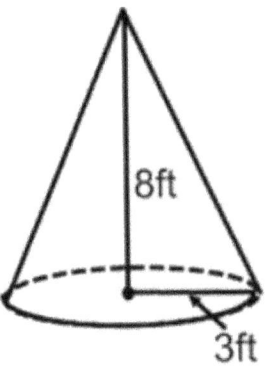

8ft

3ft

12. **Select the equation you would use to find the volume of the sphere pictured.**

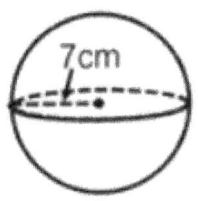

7cm

Ⓐ $4 \times \pi \times 7^2$

Ⓑ $\frac{4}{3} \times \pi \times 7^3$

Ⓒ $\frac{4}{3} \times \pi \times 7^2$

Did You Check Your Score?

YES	NO
Record your score below: Score (%): _____ Date: _____	► Scan the QR code or Visit *lumoslearning.com/a/8m029* ► Submit your answers using the *Online Answer Sheet*. ► Get your Scores & Detailed Explanations.

End of Geometry

Chapter 5
Statistics and Probability

DIRECTIONS

Do **NOT** write your answers in this book. **OPEN** the Online Answer Sheet by Scanning the **QR Code** or Visit **lumoslearning.com/a/8m030**

Lesson 1: Interpreting Data Tables & Scatter Plots

1. If a scatter plot has a line of best fit that decreases from left to right, which of the following terms describes the association?

   Ⓐ Positive association
   Ⓑ Negative association
   Ⓒ Constant association
   Ⓓ Nonlinear association

2. If a scatter plot has a line of best fit that increases from left to right, which of the following terms describes the association?

   Ⓐ Positive association
   Ⓑ Negative association
   Ⓒ Constant association
   Ⓓ Nonlinear association

3. Which of the following scatter plots is the best example of a linear association?

   Ⓐ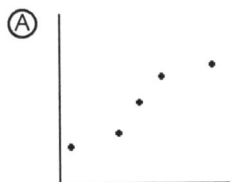
   Ⓒ

   Ⓑ
   Ⓓ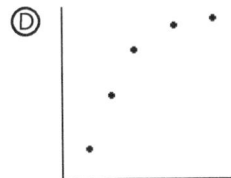

4. Data for 9 kids' History and English grades are made available in the chart. What is the association between the History and English grades?

Kids	1	2	3	4	5	6	7	8	9
History	63	49	84	33	55	23	71	62	41
English	67	69	82	32	59	26	73	62	39

Ⓐ Positive association
Ⓑ Negative association
Ⓒ Nonlinear association
Ⓓ Constant association

5. Data for 9 kids' History grades and the distance they live from school are made available in the chart. What is the association between these two categories?

Kids	1	2	3	4	5	6	7	8	9
History	63	49	84	33	55	23	71	62	41
Distance from School (miles)	.5	7	3	4	5	2	3	6	9

Ⓐ No association
Ⓑ Positive association
Ⓒ Negative association
Ⓓ Constant association

6. Data for 9 kids' Math and Science grades are made available in the chart. What is the association between the Math and Science grades?

Kids	1	2	3	4	5	6	7	8	9
Science	63	49	84	33	55	23	71	62	41
Math	67	69	82	32	59	26	73	62	39

Ⓐ Positive association
Ⓑ No association
Ⓒ Constant association
Ⓓ Negative association

7. **Which of the scatter plots below is the best example of positive association?**

Ⓐ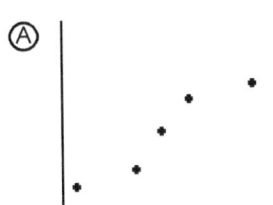

Ⓑ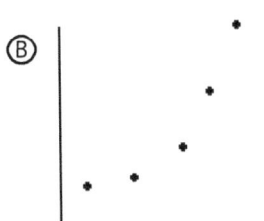

Ⓒ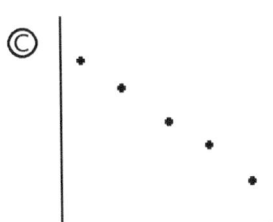

Ⓓ

8. **150 students were surveyed and asked whether they played a sport and whether they played a musical instrument. The results are shown in the table below.**

	Plays an Instrument	Does not Plays an Instrument
Plays a Sport	60	30
Does not Plays a Sport	10	50

What percent of the 150 students play a sport and also do not play an instrument?

Ⓐ **20%**
Ⓑ **33%**
Ⓒ **40%**
Ⓓ **50%**

9. 150 students were surveyed and asked whether they played a sport and whether they played a musical instrument. The results are shown in the table below.

	Plays an Instrument	Does not Plays an Instrument
Plays a Sport	60	30
Does not Plays a Sport	10	50

Which of the following statements is NOT supported by the data?

Ⓐ A randomly chosen student who plays a sport is 2 times as likely to play an instrument as to not play an instrument.
Ⓑ A randomly chosen student who does not play an instrument is 2 times as likely to not play a sport as to play a sport.
Ⓒ A randomly chosen student who does not play a sport is 5 times as likely to not play an instrument as to play an instrument.
Ⓓ A randomly chosen student who plays an instrument is 6 times as likely to play a sport as to not play a sport.

10. 150 students were surveyed and asked whether they played a sport and whether they played a musical instrument. The results are shown in the table below.

	Plays an Instrument	Does not Plays an Instrument
Plays a Sport	60	30
Does not Plays a Sport	10	50

Which two sections add up to just over half of the number of students surveyed?

Ⓐ The two sections that do not play an instrument.
Ⓑ The two sections that do not play a sport.
Ⓒ The two sections that play an instrument.
Ⓓ The two sections that play a sport.

11. Match the data with the correct association.

	POSITIVE ASSOCIATION	NEGATIVE ASSOCIATION	NO ASSOCIATION
The population survey data for 5 years shows the number of goldfish and star fish. Describe the association between the population of goldfish and star fish. <table><tr><td>YEAR</td><td>1</td><td>2</td><td>3</td><td>4</td><td>5</td></tr><tr><td>GOLDFISH</td><td>13</td><td>18</td><td>19</td><td>20</td><td>25</td></tr><tr><td>STARFISH</td><td>30</td><td>25</td><td>20</td><td>15</td><td>12</td></tr></table>	○	○	○
Below is data for 5 years showing Jonny's and Jack's weight in kg. Describe the association between the weight of Jonny and Jack. <table><tr><td>YEAR</td><td>1</td><td>2</td><td>3</td><td>4</td><td>5</td></tr><tr><td>JONNY</td><td>30</td><td>40</td><td>42</td><td>49</td><td>52</td></tr><tr><td>JACK</td><td>30</td><td>35</td><td>40</td><td>45</td><td>50</td></tr></table>	○	○	○
The data for 5 days shows the sale of watermelon and potatoes. Describe the association between the sale of watermelon and potatoes. <table><tr><td>DAYS</td><td>1</td><td>2</td><td>3</td><td>4</td><td>5</td></tr><tr><td>WATERMELON</td><td>70</td><td>26</td><td>60</td><td>19</td><td>70</td></tr><tr><td>POTATO</td><td>10</td><td>40</td><td>15</td><td>80</td><td>22</td></tr></table>	○	○	○

Did You Check Your Score?

YES

Record your score below:

Score (%): _____

Date: _____

NO

► Scan the QR code or Visit *lumoslearning.com/a/8m030*

► Submit your answers using the *Online Answer Sheet*.

► Get your Scores & Detailed Explanations.

DIRECTIONS

Do **NOT** write your answers in this book. **OPEN** the Online Answer Sheet by Scanning the **QR Code** or Visit **lumoslearning.com/a/8m031**

Chapter 5 → Lesson 2: Scatter Plots, Line of Best Fit

1.

Which of the following best describes the points in this scatter plot?

- Ⓐ Increasing Linear
- Ⓑ Decreasing Linear
- Ⓒ Constant Linear
- Ⓓ None of these

2.

Which of the following best describes the points in this scatter plot?

- Ⓐ Increasing Linear
- Ⓑ Decreasing Linear
- Ⓒ Constant Linear
- Ⓓ None of these

3.

Which of the following lines best approximates the data in the scatter plot shown above?

Ⓐ

Ⓑ

Ⓒ

Ⓓ None of these; the data do not appear to be related linearly.

4. Which scatter plot represents a positive linear association?

Ⓐ

Ⓑ

Ⓒ

Ⓓ

5. Which scatter plot represents a negative linear association?

Ⓐ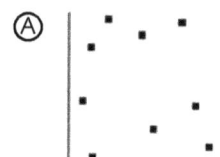

Ⓑ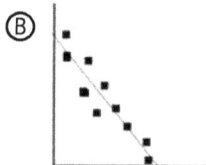

Ⓒ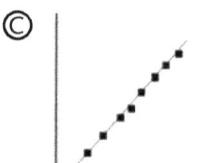

Ⓓ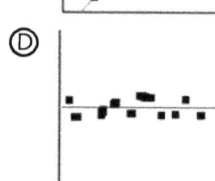

6. Which scatter plot represents no association?

Ⓐ

Ⓑ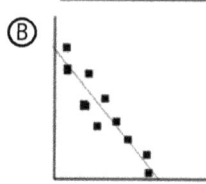

Ⓒ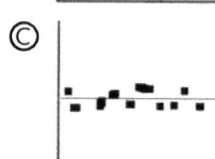

Ⓓ

7. Which scatter plot represents a constant association?

Ⓐ

Ⓑ

Ⓒ

Ⓓ

8. The graph of this data set would most resemble which of the following graphs?

x	1	2	3	4	5	6	7
y	2	3	4	5	6	7	8

Ⓐ

Ⓑ

Ⓒ

Ⓓ

9. The graph of this data would most resemble which of the following graphs?

x	1	2	3	4	5	6	7
y	4	4	4	4	4	4	4

Ⓐ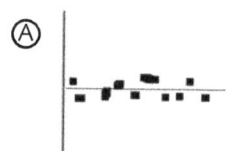

Ⓑ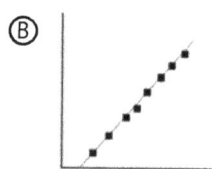

Ⓒ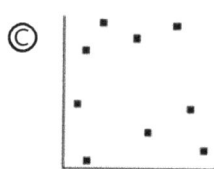

Ⓓ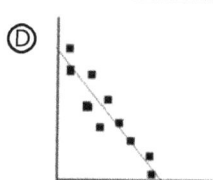

10. Typically air temperature decreases through the night between midnight and 6:00 am. This is an example of what type of association?

Ⓐ constant association
Ⓑ positive linear association
Ⓒ negative linear association
Ⓓ no association

LumosLearning.com

11. Match which line would be the best fit to describe the data pictured.

Figure - 1

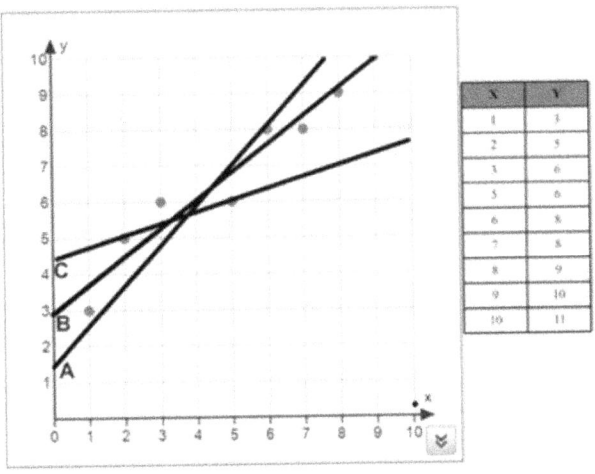

Figure - 2

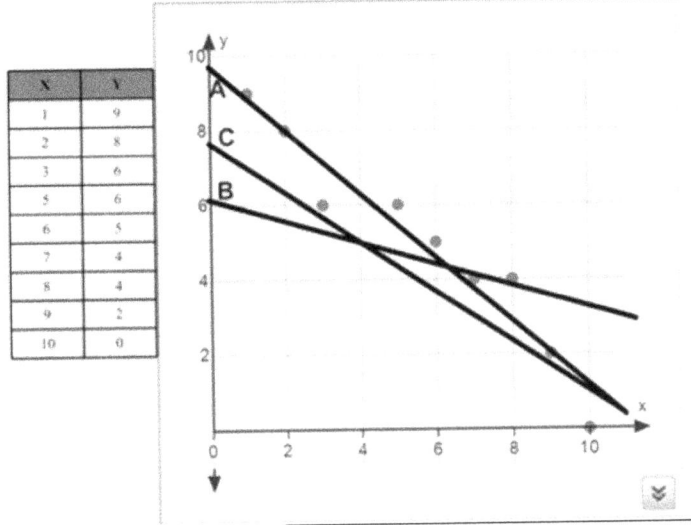

	A	B	C
Figure - 1	○	○	○
Figure - 2	○	○	○

12. Write the prediction equation for this graph using the two labeled points. Leave as fractions.

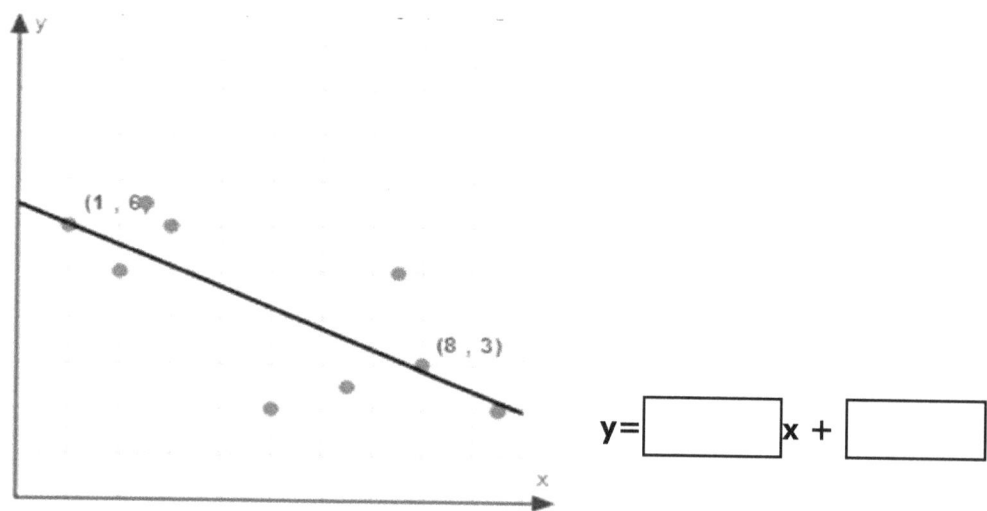

(1 , 6)

(8 , 3)

y= [   ]x + [   ]

13. Cathy wanted to know what kind of shows the 8th grade class preferred – dramas or comedies. 55 students said they liked comedies and not dramas. 25 students liked both dramas and comedies. There were 41 students who did not like dramas nor comedies. Complete a two way table using the information given.

	Doesn't Like Dramas	Likes Dramas	Total
Doesn't Like Comedies	41		97
Likes Comedies		25	
Total	96		177

Did You Check Your Score?

YES	NO
Record your score below: Score (%): _____ Date: _____	► Scan the QR code or Visit *lumoslearning.com/a/8m031* ► Submit your answers using the *Online Answer Sheet*. ► Get your Scores & Detailed Explanations.

DIRECTIONS

Do **NOT** write your answers in this book. **OPEN** the Online Answer Sheet by Scanning the **QR Code** or Visit **lumoslearning.com/a/8m032**

Chapter 5 → Lesson 3: Analyzing Linear Scatter Plots

1.

Which of the following scatter plots below demonstrates the same type of data correlation as the one shown above?

Ⓐ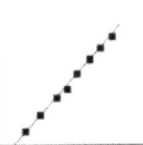

Ⓒ

Ⓑ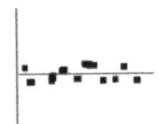

Ⓓ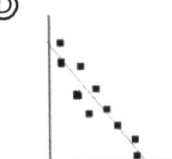

2.

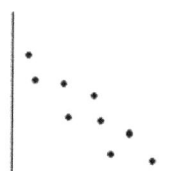

Which of the following lines most accurately models the points in this scatter plot?

Ⓐ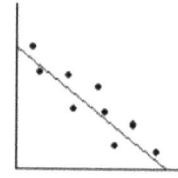

Ⓒ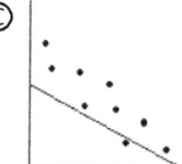

Ⓑ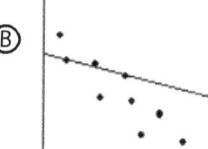

Ⓓ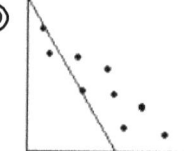

3.

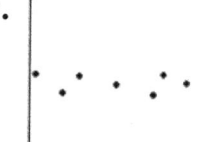

Which of the following lines most accurately models the points in this scatter plot?

Ⓐ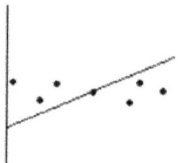

Ⓒ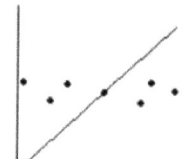

Ⓑ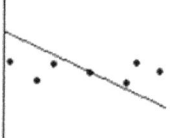

Ⓓ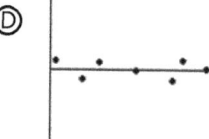

4.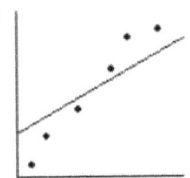

Which of the following lines most accurately models the points in this scatter plot?

Ⓐ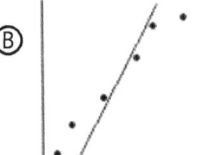

Ⓒ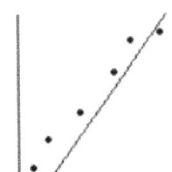

Ⓑ

Ⓓ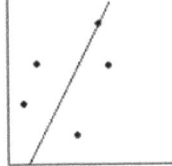

5. The four scatter plots below have all been modeled by the same line. Which scatter plot has the strongest association?

Ⓐ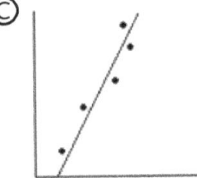

Ⓒ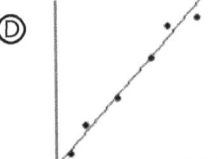

Ⓑ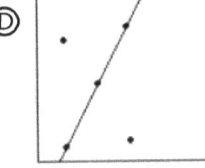

Ⓓ

6. The four scatter plots shown below have four points in common, and each scatter plot has a different fifth point. Which scatter plot's fifth point is an outlier?

Ⓐ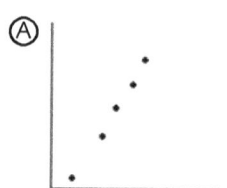

Ⓒ

Ⓑ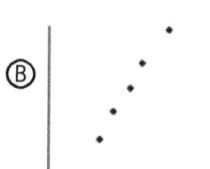

Ⓓ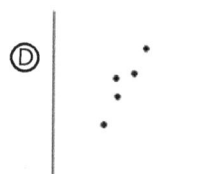

7. The four scatter plots shown below have four points in common, and each scatter plot has a different fifth point. Which scatter plot's fifth point is NOT an outlier?

Ⓐ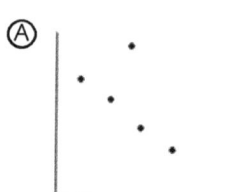

Ⓒ

Ⓑ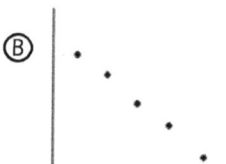

Ⓓ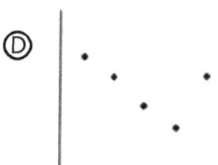

LumosLearning.com

8. The figure below shows a scatter plot relating the length of a bean plant, in centimeters, to the number of days since it was planted. The slope of the associated line is 2. Which of the following correctly interprets the slope?

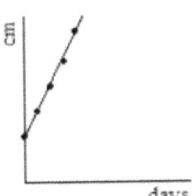

Ⓐ The bean plant grows approximately 1 cm every 2 days.
Ⓑ The bean plant grows approximately 2 cm each day.
Ⓒ The bean plant was 2 cm long when it was planted.
Ⓓ The bean plant approximately doubles in length each day.

9. The figure below shows a scatter plot relating the cost of a ride in a taxicab, in dollars, to the number of miles traveled. The slope of the associated line is 0.5. Which of the following correctly interprets the slope?

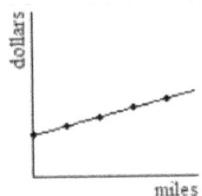

Ⓐ For each additional mile traveled, the cost of the ride increases by 50 cents.
Ⓑ For each additional half of a mile traveled, the cost of the ride increases by 1 dollar.
Ⓒ The initial cost of the ride, before the taxi has traveled any distance, is 50 cents.
Ⓓ The first half of a mile does not cost anything.

10. The figure below shows a scatter plot relating the temperature in a school's parking lot, in degrees Fahrenheit, to the number of hours past noon. The slope of the associated line is -3. Which of the following correctly interprets the slope?

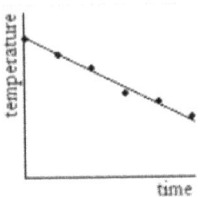

Ⓐ The temperature at noon was -3 degrees Fahrenheit.
Ⓑ The temperature decreased until it reached -3 degrees Fahrenheit.
Ⓒ The temperature decreased an average of 1 degree Fahrenheit every 3 hours.
Ⓓ The temperature decreased an average of 3 degrees Fahrenheit per hour.

11. Match the correct vocab term with the correct definition.

	Linear	Negative Association	Line of Best Fit	Prediction Equation
A line on a graph showing the general direction that a group of points seem to be heading	○	○	○	○
A graph that is represented by a straight line	○	○	○	○
The equation of a line that can predict outcomes using given data	○	○	○	○
A correlation of points that is linear with a negative slope	○	○	○	○

12. Write the equation of the best fit line for this scatter plot using the 2 ordered pairs given. Write your answer in the box given below.

Did You Check Your Score?

YES	NO
Record your score below:	► Scan the QR code or Visit *lumoslearning.com/a/8m032*
Score (%): _____	► Submit your answers using the *Online Answer Sheet*.
Date: _____	► Get your Scores & Detailed Explanations.

DIRECTIONS

Do **NOT** write your answers in this book. **OPEN** the Online Answer Sheet by Scanning the **QR Code** or Visit **lumoslearning.com/a/8m033**

Chapter 5 → Lesson 4: Relatable Data Frequency

1. 50 people were asked whether they were wearing jeans and whether they were wearing sneakers. The results are shown in the table below.
   What percent of the people who wore sneakers were also wearing jeans?

	Jeans	No Jeans
Sneakers	15	10
No Sneakers	5	20

Ⓐ 15%
Ⓑ 30%
Ⓒ 60%
Ⓓ 75%

2. 50 people were asked whether they were wearing jeans and whether they were wearing sneakers. The results are shown in the table below.
   What percent of the people who wore jeans were also wearing sneakers?

	Jeans	No Jeans
Sneakers	15	10
No Sneakers	5	20

Ⓐ 15%
Ⓑ 30%
Ⓒ 60%
Ⓓ 75%

3. 50 people were asked whether they were wearing jeans and whether they were wearing sneakers. The results are shown in the table below.
What percent of the people who did NOT wear sneakers were wearing jeans?

	Jeans	No Jeans
Sneakers	15	10
No Sneakers	5	20

Ⓐ 5%
Ⓑ 20%
Ⓒ 25%
Ⓓ 40%

4.

	Fertilizer	No Fertilizer
Lived	200	600
Died	50	150

Out of 1,000 plants, some were given a new fertilizer and the rest were given no fertilizer. Some of the plants lived and some of them died, as shown in the table above. Which of the following statements is supported by the data?

Ⓐ Fertilized plants died at a higher rate than unfertilized plants did.
Ⓑ Fertilized plants and unfertilized plants died at the same rate.
Ⓒ Fertilized plants died at a lower rate than unfertilized plants died.
Ⓓ None of the above statements can be supported by the data.

5.

	Windy	Not Windy
Sunny	5	15
Cloudy	4	6

The weather was observed for 30 days; each day was classified as sunny or cloudy, and also classified as windy or not windy. The results are shown in the table above. Which of the following statements is NOT supported by the data?

Ⓐ 25% of the sunny days were also windy.
Ⓑ 30% of the days were windy.
Ⓒ 40% of the cloudy days were also windy.
Ⓓ 50% of the windy days were also sunny.

LumosLearning.com

6.

	Jeans	No Jeans
Sneakers	15	10
No Sneakers	5	20

50 people were asked whether they were wearing jeans and whether they were wearing sneakers. The results are shown in the table above.
What fraction of the people who wore sneakers were NOT wearing jeans?

Ⓐ $\frac{1}{5}$

Ⓑ $\frac{2}{5}$

Ⓒ $\frac{3}{10}$

Ⓓ $\frac{3}{4}$

7.

	Jeans	No Jeans
Sneakers	15	10
No Sneakers	5	20

50 people were asked whether they were wearing jeans and whether they were wearing sneakers. The results are shown in the table above. Which of the following statements is NOT supported by the data?

Ⓐ A randomly chosen person who is wearing sneakers is equally as likely to be wearing jeans as not wearing jeans.

Ⓑ A randomly chosen person who is not wearing jeans is 2 times as likely to be not wearing sneakers as wearing sneakers.

Ⓒ A randomly chosen person who is wearing jeans is 3 times as likely to be wearing sneakers as not wearing sneakers.

Ⓓ A randomly chosen person who is not wearing sneakers is 4 times as likely to be not wearing jeans as wearing jeans.

8.

	Enjoys Crosswords	Does not Enjoy Crosswords
Enjoys Sudoku	30	20
Does not Enjoy Sudoku	40	10

100 people were asked whether they enjoy crossword puzzles and whether they enjoy sudoku number puzzles. The results are shown in the table above.

What percent of all 100 people enjoy sudoku?

Ⓐ 20%
Ⓑ 30%
Ⓒ 50%
Ⓓ 60%

9.

	Enjoys Crosswords	Does not Enjoy Crosswords
Enjoys Sudoku	30	20
Does not Enjoy Sudoku	40	10

100 people were asked whether they enjoy crossword puzzles and whether they enjoy sudoku number puzzles. The results are shown in the table above.
Which of the following statements is NOT supported by the data?

Ⓐ A randomly chosen person who enjoys sudoku is more likely to enjoy crosswords than to not enjoy crosswords.
Ⓑ A randomly chosen person who does not enjoy sudoku is more likely to enjoy crosswords than to not enjoy crosswords.
Ⓒ A randomly chosen person who enjoys crosswords is more likely to enjoy sudoku than to not enjoy sudoku.
Ⓓ A randomly chosen person who does not enjoy crosswords is more likely to enjoy sudoku than to not enjoy sudoku.

10.

Preferred Sports

	Volleyball	Basketball	Softball
Boys	5	30	15
Girls	30	5	15

Out of those students who preferred volleyball, about what percent were girls?

Ⓐ 15%
Ⓑ 35%
Ⓒ 85%
Ⓓ 100%

11. Using the data from the table below match the answers to the questions about the table. June surveyed the 7th and 8th grades to see which class they liked better, math or English. The results are shown in the two-way table below.

	Math	English
7th grade	78	67
8th grade	86	45

	131	.60	145	.66
Total number of 7th graders surveyed	○	○	○	○
Total number of 8th graders surveyed	○	○	○	○
The relative frequency of 7th grade students that chose English to all students that chose English	○	○	○	○
The relative frequency of 8th grader students that chose Math to the total number of 8th graders	○	○	○	○

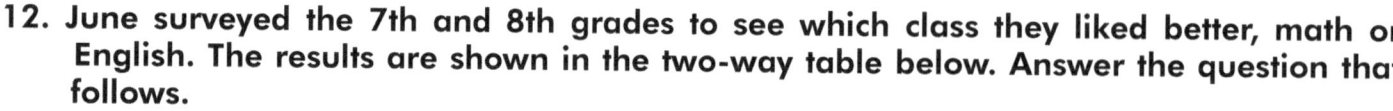

12. June surveyed the 7th and 8th grades to see which class they liked better, math or English. The results are shown in the two-way table below. Answer the question that follows.

	Math	English
7ᵗʰ grade	78	67
8ᵗʰ grade	86	45

The relative frequency of 7th grade students that chose math to all 7th grade students is _____.

13. Sam surveyed his classmates to find out if they played a sport after school or practiced in the school band. Fourteen of his classmates played a sport. Of those 14, only 5 participated in the school band. Eight students played in the school band. There were ten students who did not play a sport nor participate in the school band. Fill in the blanks in the table given below.

	Plays a Sport	Doesn't Play a Sport	Total
Plays in Band	5		8
Not a Band		10	
Total		13	27

Did You Check Your Score?

YES	NO
Record your score below: Score (%): _____ Date: _____	► Scan the QR code or Visit **lumoslearning.com/a/8m033** ► Submit your answers using the *Online Answer Sheet*. ► Get your Scores & Detailed Explanations.

End of Statistics and Probability

Notes

Wisconsin Forward Exam Test Prep: 8th Grade Math Practice Workbook and Full-length Online Assessments: WFE Study Guide

Contributing Editor - Nicole Fernandez
Contributing Editor - Nancy Chang
Contributing Editor - Greg Applegate
Executive Producer - Mukunda Krishnaswamy
Program Director - Anirudh Agarwal
Designer and Illustrator - Sowmya R.

COPYRIGHT ©2024 by Lumos Information Services, LLC. ALL RIGHTS RESERVED. No portion of this book may be reproduced mechanically, electronically or by any other means, including photocopying, recording, taping, Web Distribution or Information Storage and Retrieval systems, without prior written permission of the Publisher, Lumos Information Services, LLC.

Wisconsin Department of Education is not affiliated to Lumos Learning. Wisconsin Department of Education, was not involved in the production of, and does not endorse these products or this site.

ISBN 13: 979-8897130139

Printed in the United States of America

CONTACT INFORMATION

LUMOS INFORMATION SERVICES, LLC

 PO Box 1575, Piscataway, NJ 08855-1575
 www.LumosLearning.com

Email: support@lumoslearning.com
Tel: (732) 384-0146
Fax: (866) 283-6471

Lumos Learning
Step Up Your Skills

What if I buy more than one Lumos tedBook?

Step 1 → **Visit the link given below and login to your parent/teacher account**

www.lumoslearning.com

Step 2 → Go to the **"My tedBooks"** section and place the book access code and submit (See the first page for access code).

Step 3 → **Add the new book**

To add the new book for a registered student, choose the **'Student'** button and click on submit.

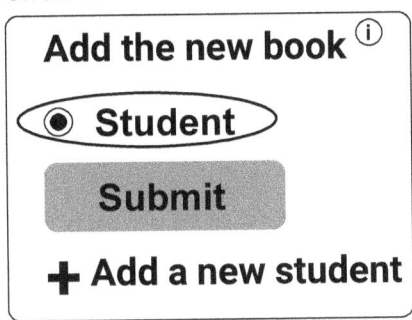

To add the new book for a new student, choose the **'Add New Student'** button and complete the student registration.

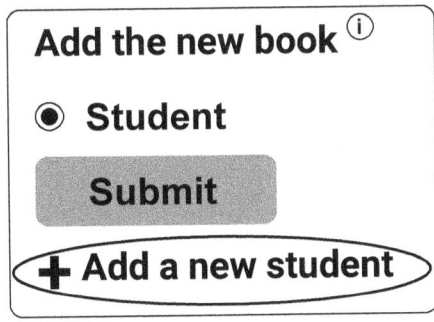

Also Available

Wisconsin Forward Exam Test Prep: Grade 3 English Language Arts Literacy (ELA)
Practice Workbook and Full-length Online Assessments: WFE Study Guide

Grade **8**

Lumos Learning
Step Up Your Skills

WISCONSIN

ENGLISH
LANGUAGE ARTS LITERACY

WFE Practice

ONLINE

2 Full-Length Practice Tests

Personalized Study Plan & Resources

Automated Scoring and Instant Feedback

ELA Strands   Literature • Informational Text • Language

www.ingramcontent.com/pod-product-compliance
Lightning Source LLC
Chambersburg PA
CBHW041115120626
46547CB00019B/2724